No. 1965
$15.95

CELLULAR TELEPHONES
A LAYMAN'S GUIDE

STUART CRUMP, JR.
WITH AFTERWORD BY JOHN NAISBITT

TAB BOOKS Inc.
Blue Ridge Summit, PA 17214

*To my wife and my parents,
who made it all possible.*

FIRST EDITION

FIRST PRINTING

Copyright © 1985 by TAB BOOKS Inc.

Printed in the United States of America

Reproduction or publication of the content in any manner, without express permission of the publisher, is prohibited. No liability is assumed with respect to the use of the information herein.

Library of Congress Cataloging in Publication Data

Crump, Stuart.
    Cellular telephones.

    Includes index.
    1. Cellular radio. I. Title.
TK6570.M6C76   1985   384.5'3   85-16850
ISBN 0-8306-0965-2
ISBN 0-8306-1965-8 (pbk.)

Front cover photograph courtesy of Motorola Inc., Communications Sector, Schaumburg, IL.

Table of Contents

INTRODUCTION *by Martin Cooper* — ix

FOREWORD — xi

1. **YOU CAN AFFORD A CAR TELEPHONE** — 1
   - What is Cellular Telephone? — 2
   - Cellular Telephones: An Expensive Luxury or a Bargain? — 3
   - Why Do You Need a Car Telephone? — 3
   - Alternatives to Cellular — 4
   - Not for the Rich Only — 4
   - You CAN Afford a Car Phone — 5

2. **WHAT IS CELLULAR TELEPHONE?** — 6
   - How Cellular Works—in Non-Technical Terms

3. **THE POLITICS OF CELLULAR TELEPHONE** — 15
   A Basic Overview of the Regulatory Struggle to Bring Cellular On-Line
   - The History of Cellular Telephone — 16
   - Experimental Systems — 17

4. **HOW MUCH IS YOUR TIME WORTH?** *by Dr. Larry Baker* — 20
   Compared to the Value of Your Time, a Car Phone is a Bargain, Not a Burden
   - How to Determine the Value of Your Time to Your Company — 21
   - Not Everyone Needs One — 22
   - Excellent for Sales Personnel — 23
   - Analyzing Your Mobile Communication Needs — 24
   - How to Calculate the Value of Your Time — 27
   - How Much is Your Time Worth? — 32
   - A Sample Case — 32
   - Examples of How Valuable a Car Phone Can Be — 34
   - Personal Use—Mobile Telephones — 40
   - Another Way to Look at the Value of Your Time — 41
   - Cellular Telephone: An Expensive Luxury or a Bargain? — 42
   - Less Expensive than Opening and Equipping an Office — 43
   - Conclusion: A Cellular Phone Can be a Bargain — 43

5. **HOW OTHERS ARE USING THEIR CELLULAR TELEPHONES** — 44
   - Contractor Doubled His Business — 46
   - Lawyer Increases His Income — 46
   - A $2,000 Per Month Phone Bill — 46
   - Other Examples — 47
   - Time is Money — 47

6. **WHY YOU SHOULD PURCHASE YOUR CAR PHONE NOW** — 49
   If You Need a Car Phone Now, Don't Wait "Until Prices Drop"
   - Benefit from Your Car Phone Today — 52

7.	**HOW MUCH DOES IT COST?**	**54**
	The Three Parts of Your Cellular Telephone Bill	54
	Where to Buy Your Car Phone	58
8.	**HOW TO SELECT YOUR CAR PHONE**	**61**
	A Smorgasborg of Features to Choose From	
9.	**MOUNTING THE PHONE IN YOUR CAR**	**73**
	Convenience and Safety are Prime Concerns	
10.	**HOW SAFE IS YOUR CAR PHONE?**	**77**
	If You Exercise Caution, Your Car Phone Will Offer You Many Years of Safe Use	
11.	**HOW TO GET FREE SERVICE— AND OTHER CONSIDERATIONS**	**80**
	Leasing Your Car Phone	81
	How to Rent a Cellular Phone	81
	Buy Your Car Phone at the Same Time You Buy Your Car	82
	O, Give Me a Home Where the Mobilephones Roam	82
	...But Not in the Air	83
	How to Get Free Cellular Service	85
12.	**HOW TO GET THE MAXIMUM STATUS VALUE OUT OF YOUR NEW CAR PHONE** *by Sue Easton*	**87**
	How to Let Them Know	88
	One-Upmanship	91
13.	**THE CAR PHONE vs. THE PORTABLE PHONE**	**92**
	Which is the Better Choice for You?	
	Solving the Battery Problem	94
	Join the Beep Generation	94
	Move the Decimal Point	95
	Do You Need a Portable Phone?	96
	Advantages and Disadvantages	97
	The Transportable	98
	The Briefcase Phone	98
	Making the Case for the Briefcase Phone *by Benn Kobb*	99
14.	**HOW TO MAKE THE MOST EFFICIENT USE OF YOUR CAR PHONE**	**105**
	Incoming and Outgoing Calls	106
	How to Make Your Car Phone an Extension of Your PBX	106
	Call Diverting Increases Your Mobility	107
	After Hours and Weekend Use	107
15.	**SECURITY: HOW PRIVATE ARE CELLULAR PHONES?**	**109**
	How Secure are Telephone Calls Today?	111
	The Digital Solution	111

16. THE ALTERNATIVES TO CELLULAR		**113**
	Paging: Things that Go Beep in the Night	114
	Pagers are for Everyone	117
	Telephone Answering Machines	118
	Voice Mail: A Computerized "Answering Machine" *by Elaine Lussier*	118
	Electronic Mail	121
	Conventional Two-Way Radio	122
	Specialized Mobile Radio	122
	Wireless Data Communications	123
17. THE FUTURE OF MOBILE COMMUNICATIONS		**125**
	A Bright Future	127
AFTERWORD: "Cellular: The Beginning of a Revolution" The Social and Political Implications of Person-to-Person Global Telecommunications will Play a Major Role in Your Future *by John Naisbitt*		**129**
	Smaller and Smaller	131
	Pocket-size Phones	131
	Breaking National Barriers	132
	Bypassing Governments	132
	Love at First Dial	133
APPENDIX: How to Make a Safe Call from Your Car		**135**
THE AUTHORS		**139**
PHOTO CREDITS		**143**
INDEX		**145**

Introduction

Cellular radiotelephone is not the first multi-billion dollar industry to spring from the minds of our visionaries and technocrats, but the impact of this new industry upon an important segment of our communicating society will be just as profound as others such as the invention of the steam engine and the dynamo.

In 1983, the total revenues for cellular operation in the U.S. were a few hundred thousand dollars. By contrast, research and development costs to develop the service had exceeded $30 million.

By 1993, total revenues for equipment suppliers, service vendors and cellular operator billings are expected to exceed $10 billion.

The addition of $10 billion of products and services to our economy is significant enough to make an indelible mark on our history, but the economic and social impact of cellular radio will be far greater than $10 billion. That's what this book is all about.

People will subscribe to cellular service only if they expect a return on their investments. Most of the tens of thousands of these people who became cellular subscribers each month in the last half of 1984 were convinced that their investment was sound—that cellular telephones were useful and desirable.

These early buyers—these pioneers—will discover that their cellular phones will start paying off immediately. They will also discover that, as they use their phones more and as the variety and number of ancillary service grows, the cellular phone will change their business style, improve their productivity and make them generally more effective—all beyond their most optimistic expectations.

A wild claim?

Consider that the common justification for becoming a cellular subscriber is that hour or so of commuting time can be used to conduct business calls. Commuting time, however, is not where the most exciting opportunity lies.

I predict that the subscribers who expect to make their calls in the morning or evening will discover that they can spend more time—productively—out in the field, in the real world.

Contrast this new mobility with your current, non-mobile business style. Today, you are tied to your desk so your colleagues will know where to reach you because *that* is where your phone is—wired to the wall.

The Renaissance Age of Cellular will not happen instantly. As we learned with the radio paging, the growth of any new industry takes time. Some professions will adapt to cellular more quickly than others. The cellular subscriber of 1989 will be different from the 1984 subscriber. He will be skillful in the use of the tool, more comfortable with the technology and more mobile.

This book is a fine first attempt to accelerate the process of creating the cellular subscribers of the future. The book itself, insofar as it will help you avoid some pitfalls and teach you to get more out of your cellular phone, is a productivity improver.

Read it. Profit from it. Your whole lifestyle is about to change—for the better.

> Martin Cooper, Chairman
> Cellular Business Systems Inc.
> Chicago, Illinois

Foreword

The cellular revolution has begun.

The cellular telephone is more than just another high-priced electronic toy for the wealthy. It is a time-management tool—and it is for *everyone*..

It is such an important new tool to help you manage your day and your life and increase your productivity that—contrary to what you may have heard—the cellular telephone is an inexpensive *bargain,* not an overpriced luxury.

If you find that statement hard to believe, look at it this way:

How much would you be willing to spend to add an extra hour or two to your busy day? How much is your time worth? It is worth far more than you realize—and compared to the value of your time, the car phone is one of the greatest bargains to come along in years. (We will explore that question in depth in Chapter 4, which is perhaps the most important chapter in this book.)

Time is the one resource we never have enough of.

Have you ever taken a time management course or read a book on time management, hoping that you might discover that magic secret of how to create more time? Time is, after all, the most valuable commodity of all.

Most time management methods require you to totally change your lifestyle and account for every minute to successfully implement their techniques.

If you heard of a new tool that would add that hour or two to your day but required little or no change in your lifestyle, you probably would not even ask how much it would cost. Your first question would be, "Where can I get it?"

The typewriter, the word processor, the telephone, the photocopy machine, the automobile, the airplane and other such tools increase our productivity—our ability to produce more in less time.

The *cellular telephone* is another new and revolutionary tool that increases the productive time of today's business person.

This book will tell you all you need to know about the cellular telephone—what it is, how it works, how to select your car phone and how to use it to increase productivity—and profit—in your business and in your personal life.

When you finish reading this book and begin using your own cellphone, you will understand what we mean when we say that the cellular telephone is the first step in the *personal communications* revolution that will change the way we live and work.

*　　　*　　　*

Author's acknowledgment: I am indebted to the speakers at the American Management Association's session, held in May 1984, on "Mobile Communications Technologies: Facts and Formulas for Dollar-Wise Decision Making." They were: Al Ehrlich, section chief, marketing communications, AT&T Network Systems; Bob Martin, vice president, marketing, NYNEX Mobile Communications; Charles Hough, president, Communications Systems Integration of America Inc.; and David Post, chairman, PageAmerica Communications Inc. We are all indebted to Charlotte Cooney of the AMA, who spawned the idea for the seminar that inspired this book, supported it and helped put it together.

And, of course, I couldn't have done it without the help of all members of our office staff, especially Lucille Jackson, who sat through the tedious process of transcribing this book off my dictation tapes, and Lee Greathouse, who must have lost 10 pounds editing my copy.

When you finish reading this book you will understand what we mean when we say, "Cell the World!"

Stuart Crump Jr., Editor and Publisher
Personal Communications Magazine
Fairfax, Virginia

Personal Communications Report, the monthly telecommunications newsletter for consumers, has a subscription price of $24 per year. *Personal Communications Magazine*, which began publication in May 1983, is $25 per year. Both are published by FutureComm Publications Inc., 4005 Williamsburg Ct., Fairfax, VA 22032, 703/352-1200.

Chapter 1
You CAN Afford a Car Telephone

Car Telephones are No Longer for the Rich Only!
The New "Cellular" Technology Brings
Them within Everyone's Budget

YOUR CAR PHONE is always there when you need it—for business, personal and even emergency use. This unit is made by NovAtel.

What image goes through your mind when you hear the words *car telephone?*

Most likely, you envision a rich industrialist or company president wearing a three-piece, pin-striped suit sitting in the back seat of his chauffeur-driven limousine, closing a multimillion-dollar deal as he is driven across town.

Or perhaps you envision a character on "Charlie's Angels" or "Hart to Hart" or some other television private-eye show gathering information via car phone while barreling down the highway trying to solve yet another baffling crime before the next commercial.

In other words, the traditional image conveyed by the words *car telephone* is that such devices are only for the rich and prominent. At the very least, the car telephone connotes an image of wealth, status and success.

Until the advent of the cellular telephone, car telephones were, indeed, primarily a tool (perhaps even a toy) for the rich. The cellular telephone introduces us to a new era of mobile communications—an era in which anyone who wants to own his own car telephone will be able to do so within just a few years. Indeed, you may even be able to afford a car telephone *today*. This book will explain some of the ways in which you can add a telephone to your car without putting yourself in the poorhouse.

What is Cellular Telephone?

To explain cellular radio—or cellular telephone (the terms are interchangeable)—we will begin by examining traditional mobile telephones.

In the traditional car phone, a large antenna is placed in the center of the city, blanketing the entire city with the signals from the transmitter. Because the signal covers the whole city, only one person may use one channel to place one phone call at any one time—a very wasteful use of the radio portion of the electromagnetic spectrum.

Cellular telephone, on the other hand, divides a city into a number of small "cells," each with its own low-powered transmitter. Each transmitter is strong enough to serve that cell only and does not interfere with the signals in other, nonadjacent cells in the city. Cellular telephone allows the *same* frequencies to be used and *reused* several times throughout the city, thus expanding the number of mobile telephones that can be used in that city at the same time. (We will go into a more detailed explanation of how cellular works in Chapter 2.)

Traditional car-telephone service has been expensive because the scarcity of service has driven the price beyond the means of the average person. The spread of cellular service will ultimately drive the price of the car phone down, so that anyone who wants one will be able to afford it. Within a decade or so, phones in cars should be just about as common as air conditioners and am/fm-radio cassette players are in cars today.

In its initial stages of service in the United States, cellular is still priced as an expensive car telephone service—almost as expensive as the traditional car phone services it is replacing. You may recall that the first pocket calculators had price tags that made them available only to wealthy people. The first video cassette recorders were also priced out of the reach of most people.

Cellular Telephone: An Expensive Luxury or a Bargain?

Depending upon how you value your time, the cost of a cellular telephone in your car can actually be a bargain—not an overpriced luxury—especially if your job frequently requires you to travel by car.

We will get into the question of how much your time is worth and why cellular service is a bargain—not an expensive luxury—in Chapter 4.

Why Do You Need a Car Telephone?

You might wish to have your own car phone for one or more of any number of reasons. Some of these reasons are:

- **Prestige.** Car phones have a certain mystique about them. When you have a car phone, it indicates that you have "made it." If you use your car to entertain clients or other important people, you may wish to have a car phone even if you do not use it often—in order to indicate in a subtle way that you have "reached the top."

- **A business tool.** If you spend a half hour or more in your car every day or if your managerial or supervisory function is so important that the business cannot function at 100% when you are away, you should definitely consider installing a phone in your car—regardless of price. You may be able to increase your job productivity simply by adding a phone to your car, especially if you are a real estate salesperson, lawyer, traveling salesperson, journalist or repairman (such as plumber and electrician).

The car phone is a business tool that can give you as many as one or two extra hours every day that otherwise might be lost while you are stuck in traffic.

- **Home management tool.** Executives are not the only people who have to manage other people. Homemakers are constantly on the move, shuttling children between school plays, ball games and Scouts, not to mention shopping and running errands all over town. A car phone can help the homemaker who is constantly on the go maintain her mental balance in an otherwise hectic world by giving her the assurance of always being in touch and in the know.
- **Peace of mind.** When you step into your car, it's almost as if you step into another dimension. You become inaccessible to the rest of the world. What if a crisis develops that demands your attention? What if a child or a friend or a family member needs to be picked up? What if you have planned a meeting and one party is unable to keep the appointment? Small and large crises such as these can be resolved quickly and efficiently with a car telephone.
- **Attraction to the new.** If you are one of those people (as I am) who just *has* to have the latest new electronic gadget or device that comes on the market, you would not consider living another day without a car telephone of your own. If you have this particular fetish, this book will help you fulfill your desire at a price that won't break you.

Alternatives to Cellular

Cellular telephone is the hot new technology that is receiving most of the press attention, but it is by no means the only type of car telephone available. It may not even be the best for you.

Chapter 16 in this book will cover several other new (and old) technologies that offer you alternatives to cellular telephone. Some of these technologies may meet your particular mobile communications needs at a lower price than what you would have to pay for cellular telephone.

Not for the Rich Only

The point that you will need to understand is that car telephones are no longer for the rich only. They are becoming increasingly less expensive. You can probably afford one if you need one.

Car phones, like pocket calculators and video cassette recorders (VCRs), are starting out as expensive luxuries. Within a matter of time—one to 5 years, perhaps—they should come down to a price range that makes them affordable to almost anyone who wishes to own one.

Car phones can also be compared to air conditioning in cars. When it was first introduced, car air conditioning was an expensive luxury that very few people could afford. Today, not everyone has an air conditioner in his car, but anyone who can afford to buy a new car can almost always afford to buy an air conditioner. In many areas of the country, air conditioning is considered a necessity—not a luxury.

You CAN Afford a Car Phone

You no longer have to view the car phone as an expensive luxury. If you truly wish to have one, you can afford it. With a creative approach, you will find that the car phone is within your budget, no matter how small that budget may be.

You might want to consider the car telephone in the same general desire category as the car. Every teenage boy has that burning desire to own his first car just as soon as possible. Can he afford it? No. Nonetheless, he will scrimp and save and work double time in order to buy that car.

As Ohio State University Professor of Marketing Roger Blackwell put it in a recent speech: "Because of a proliferation of affluence, people in our society can afford absolutely anything they want, but the 'catch 22' is that they can't afford everything."

In other words, you have to be selective. If you truly want a car or a home or a membership at a country club or a baby or a VCR—or a car telephone—and if it is high enough on your priority list, you can find a way to afford it.

You can afford a car telephone. It's up to you!

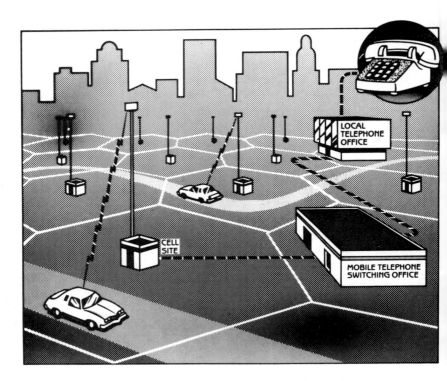

HOW CELLULAR WORKS. Cellular depends on a network of "cells" (indicated on this diagram by the hexagon-shaped segments) that cover a city. A low-powered radio transmitter and control equipment is located near the center of each cell. This cell-site equipment is connected to the "mobile telephone switching office" (MTSO), which is the gateway to the regular landline telephone network. When you place a car-phone call, the MTSO monitors the strength of your phone's signal at each of the cell sites near you. The closest site handles your call. As you move from one cell to another, your signal is automatically "handed off" to the next cell, giving you a clear, strong signal. (Drawing courtesy of Bell Atlantic Mobile Systems)

Chapter 2
What is Cellular Telephone?

A Simple, Non-Technical Explanation of Cellular in Terms that Even a Mother-in-Law Can Understand

If you were to walk up to the average "man (or woman) in the street" and ask him (her) if he (she) has ever heard of cellular, few would be able to say "yes." A few people might say something like, "That's some new type of car phone, isn't it? I know somebody who has one. It's very expensive."

Even fewer people could explain how it works.

Contrary to popular opinion, "cellular" is not a difficult concept. The technical side of cellular is not particularly difficult to understand, provided you don't attempt to get into the nitty-gritty details such as how a cell transfers a call to another cell and how a radio transmitter works.

In this chapter we will offer a simplified explanation of the technical side of cellular—explained in non-technical terms.

In the next chapter we will cover the historical and legal background of the development of cellular telephone.

If your primary interest is in how to select and use your car telephone, I urge you to skip both these chapters and begin reading with Chapter 4. You don't need to know any of this material.

Chapters 2 and 3 in this book are designed to give you enough background so that you will be familiar with the various buzz words and other concepts connected with cellular. You don't really need to know them to use your car phone, but they may prove handy to you if you need ready facility with the esoteric aspects of cellular in order to impress your boss that you know what you are talking about when you attempt to convince him to buy you a car phone.

A long time ago (we are tempted to add, "in a galaxy far, far away"), before we began publishing *Personal Communications Magazine*, I wrote an article entitled, "The Concept [of Cellular Radio] in Terms Even a Mother-in-Law Can Understand." This article, which appeared in the June 1982 issue of *Telecourier* magazine, was apparently seen by everyone in the industry.

I dashed the article off quickly in long hand (a skill I have almost forgotten since getting my first word processor a couple of months before) on the plane while flying to a convention in Hawaii in April of that year.

Despite having written it in haste, the article apparently came off in such a way that even the newest newcomer to the cellular revolution could actually understand cellular telephone after reading it.

Due to popular demand, we reprinted the article in the January 1984 issue of *Personal Communications Magazine.* It also seems to be an appropriate chapter for this book on cellular telephone. If you already know how cellular works—or if you don't care—skip the rest of this chapter.

Cellular "Radio" vs. Cellular "Telephone"

For all practical purposes, the terms "cellular *radio*" and "cellular *telephone*" are identical and may be used interchangeably.

"Cellular radio" is the term that has been used to describe the technology since it was first proposed more than a decade ago. Cellular was looked upon as an advanced form of mobile *radio.*

Now that the service is in operation, however, it has become increasingly clear that customers are using their car phones as if they are true *telephones*—which is what they are—and not as if they were traditional two-way *radios.*

Consequently, when we* write about the technology today, we call it cellular *telephone* because it offers all of the conveniences of standard telephone without the drawbacks of earlier types of mobile radio.

In the years I have been talking up cellular, the one thing that constantly amazes me is how many people have never heard of it. And even among those who *have* heard of cellular, the appalling lack of knowledge and misunderstanding is often embarrassing.

I am not talking about people like mothers-in-law and maiden aunts, who aren't supposed to keep up with such things.

I am referring to experienced professionals in the telecommunications field—the sort of people who can tell you everything there is

---

Footnote

*This is an editorial "we." I will use both "I" and "we" interchangeably in this book. I enjoy frustrating people who majored in English.

to know about a crossbar or reed switch (which I am only faintly familiar with) but who have no grasp at all of this revolutionary new mobile telephone technology.

Some of these people are all set to invest millions of dollars in a technology totally new to them and which they are only vaguely familiar with.

In fact, my mother-in-law has taken such an interest in cellular since her daughter married a guy who's a nut on the subject that I've given serious thought to letting her do my public speaking for me. She's a better speaker than I am and knows as much or more about the subject than many telephone company and radio common carrier professionals.

For example, at the Land Mobile Radio Show in Denver in March 1982, I went to booth after booth and asked, "What are your plans for the cellular market?"

The majority of booth attendants had *heard* of cellular, but the number of them who didn't know much about it was startling.

One fellow who was an engineer for a well-known two-way-radio manufacturer told me at great length about a wonderful "cellular system" his company was designing for an unnamed European country. It finally dawned on me that he was talking about an "interconnected VHF" system—something that is quite different from cellular. When I pointed out the difference to him, he replied, "That's what cellular is, isn't it?" He thought "cellular" was synonymous with "mobile phone."

At other booths I found I had to give my by-now almost-famous four-minute dissertation on cellular before I could even elicit a comment.

I am happy to report that the picture I saw at that show in 1982 has improved considerably in recent years. At the 1984 show, for example, almost every booth attendant had heard of cellular. However, there are many new people coming into this industry these days who can use a crash course in the subject. That is the main reason why we are reprinting this article in this book.

So, to those of you who know as little or less than my mother-in-law did before I married her daughter, I dedicate this chapter.

Conventional Car Telephones

When I discuss cellular, I always begin by explaining how conventional mobile telephones work—the phones that are commonly referred to as "IMTS," which stands for "Improved Mobile Telephone System (or Service)." (By contrast, the Bell System—before it was broken up—referred to its cellular system as "AMPS," or "Advanced Mobile Phone Service.")

The first thing I do is draw a big circle in the middle of the page and put a dot in the center of it (Figure 2-A).

"This is a conventional mobile phone system," I say as I draw the circle. "This circle can be anywhere from 25 to 50 or even 70 miles in radius and includes the area of a big city plus the surrounding suburbs.

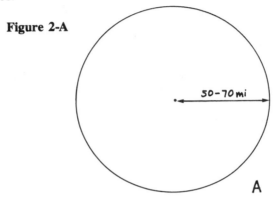

Figure 2-A

A

"In the middle of the city," I continue as I place the dot on the page, "the telephone company or radio common carrier (RCC) places a huge radio tower. Sometimes the tower goes on the top of a tall building. All the radiotelephone calls in the city go through this tower.

"In most cities," I continue, "there are only a dozen channels or so. That means only 12 people in the city can use the system at one time. At most hours during the day, too many people are trying to use too few channels. The result is that you can wait 15 or 20 minutes or even more to get a free channel so that you can place your call. The long wait can make your telephone virtually useless.

"In most areas, of course, you may not even be able to get that car telephone in the first place. Because only a few channels are available—somewhere between 12 and 40—the phone company or RCC can provide phones to only a limited number of customers. In some of the larger cities you might have to wait 5 or even 10 years or more before you will be able to get a phone for your car.

"Anyway," I continue, "let's assume for the sake of argument that you are able to obtain a car phone and that by some stroke of good fortune you have managed to grab a free channel. Your signal is picked up by this large tower in the center of the city.

The tower puts out a signal that is strong enough to reach out to the edge of the city and into the surrounding suburbs. Because this signal is so powerful, the same frequencies cannot be used in another city that is less than about 75 to 150 miles away. Otherwise, you get interference problems. Also, as you drive around the city, your signal can fade as buildings come between you and the tower," I say.

At this point, following good procedures of public speaking, I summarize what I have said thus far:

Three Main Problems

"So now you understand the three problems that conventional car phones suffer from: (1) There aren't enough channels for everyone who wants a phone, so you have to wait long periods of time—5 to 10 years—before you can get a phone, and (2) even if you do get one, you may not be able to use it because too many other people are trying to use their car phones at the same time, and (3) if you do use the phone, your signal is likely to be of poor quality because of fading, static and interference from other systems."

Most people, I find, have no trouble understanding what I have said to this point in the discussion, which generally boosts my confidence and gives me the courage to continue. I pause, take a deep breath, and go on.

"Got that?" I ask. "All right, a cellular system divides the city into small sectors—or cells—like this" (Figure 2-B).

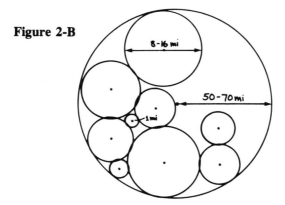

Figure 2-B

I proceed to draw a series of small, slightly overlapping circles inside the bigger circle. "Each of these circles represents a cell. That, in fact, is where the name 'cellular radio' comes from—from these 'cells.'

"At the center of each cell is placed a separate tower, coupled to a transmitter, or transceiver. This transmitter is low powered. It is strong enough to reach just about out to the edge of the cell and not much farther."

Circles within Circles

I draw a few more circles as I talk, putting a dot in the center of each of them to represent the tower and transmitter. "The low-powered transmitter puts out a signal that reaches out about *this* far—perhaps 2 to 8 miles," I say, drawing yet another circle around the dot.

"Each of these smaller circles can be anywhere from about one mile to about 8 or 10 miles or so in diameter. Suppose you are in this cell"—I point to one of the circles—"and are driving across the city while making a phone call." I draw a small car and add an arrow to show what direction it is heading (Figure 2-C).

Figure 2-C **Figure 2-D**

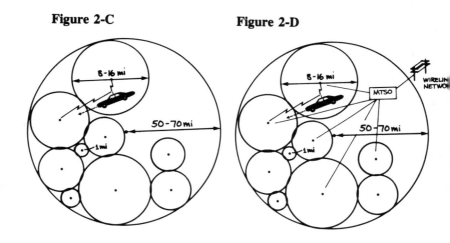

"Your car radiotelephone puts out a signal that is picked up by the tower in the cell closest to you. As you drive toward the edge of the cell, your signal will begin to fade slightly in that cell. At the same time, your signal will become stronger and stronger in the cell you are driving into.

"A central computer—called a 'mobile telephone switching office,' or 'MTSO'—constantly monitors your signal strength at each of the cell sites near your car. At some point the computer will decide, 'Hand off this call from cell A to Cell B' (Figure 2-D). And that's exactly what it does. It hands you off to the next cell.

"This *hand-off*—that's the term used to describe the process—occurs in a split second and is performed so smoothly that you generally don't even know it's happening."

Is That Clear?

Here I pause and examine the listener's face to see if he is taking it all in. This "handoff" concept can be a bit tricky the first time you hear it. When I am convinced he understands it, I continue.

"Okay, you just got handed off from one cell to the next. Now, you recall that the transmitter in cell A was quite low powered? Well, because it is, the same frequencies used in cell A can be reused over here, two or three cells away."

I put another "A" in that cell too, and do it again two or three cells away, to emphasize the reuse idea.

"The same frequencies can be used and reused many times throughout the system. By reusing the frequencies, you increase the capacity of the system. And if any one cell becomes too crowded as the number of users on the system grows, you can further increase the capacity on the system by subdividing the cells into smaller cells. This is called 'cell splitting.' "

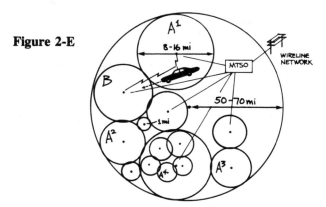

Figure 2-E

I draw even smaller circles inside one of the larger circles to demonstrate what I mean by "cell splitting" (Figure 2-E).

Often, the listener doesn't quite grasp the concept of frequency reuse, so I explain it this way:

"I live just outside Washington, D.C. We have television channels

4, 5, 7 and 9. If you go to Baltimore, about 35 miles away, you'll find that they have channels 2, 11 and 13. In Philadelphia, another 100 miles or so away, they have channels 3, 6 and 10. Then when you get to New York, another 100 miles away, you'll discover they use channels 4, 5, 7 and 9—the same channels we have in Washington.

"They can *reuse* the same channels in New York because New York is far enough away from Washington that the signals do not interfere with each other. So 'frequency reuse' is not a new concept. It has been around for years. It works the same way in cellular, except on a much smaller scale."

At this point, once again following good procedures of public speaking, I summarize:

"So now you understand the four major points that explain how cellular works: (1) The city is divided into small, low-powered *cells;* (2) Calls are routed through the nearest cell and are *handed off* from cell to cell as the car moves through the city; (3) Frequencies can be *reused* in non-adjacent cells; and (4) Cells can be divided or *split* into smaller cells to accommodate additional demand as the number of customers on the system grows."

Now You're an Expert

This is generally about as far as I go in my explanation of cellular with a first timer. I usually conclude with a comment such as, "Now you know more about cellular and how it works than half the people who are in the business itself," which, alas, is almost a true statement.

I figure that if the listener grasps what I have said, he is on his way to understanding the concept without encumbering him with more details. (My wife accuses me of dragging out explanations. She's usually right. Ask me sometime to explain to you why she's usually right, and you'll see what she means.)

Chapter 3
The Politics of Cellular Telephone

A Basic Overview of the Regulatory Struggle to Bring Cellular On-Line

HUNDREDS OF BOXES of applications (one per box) for licenses to build and operate cellular systems clog the hallways of the Federal Communications Commission in Washington as FCC employee Leonard Fields brings in yet another application.

The technical details explaining how cellular *technology* works are not all that difficult to understand. Now that you have read Chapter 2, you are an expert in the field and are ready to be drafted into the cellular army so that you, too, can bore friends and influence people as you explain to them everything they need to know about cellular but were afraid to ask.

Unfortunately, how cellular works *legally* is more difficult to understand. To paraphrase the old saying about international relations, there are only two people in the U.S. who understand the politics of cellular telephone—and they disagree.

The History of Cellular Telephone

Cellular was first proposed by scientists at Bell Laboratories in the late 1940s. The computer technology to make cellular work was developed in the 1960s.

The Federal Communications Commission (FCC) took 13 years from the time it first began discussing cellular telephone until it adopted its "final" rules in April 1981. This lengthly delay, in fact, allowed several other nations to develop and install their own cellular networks years before the technology came to the American people. Japan, for example, began operating its first cellular system in Toyko in December 1979. Cellular service began in the four Scandinavian countries in October 1981. Several Middle Eastern countries also have cellular telephone service. The system in Saudi Arabia, for example, went on the air in September 1981. Commercial service did not get under way in the U.S. until October 1983.

Under its final rules, the Commission decreed that the cellular spectrum would be divided into three parts. Of a total of 999 thirty-kHz-wide channels in the 800 MHz band, one-third, or 333 channels, would be reserved for the local telephone company in an area, one-third would be reserved for anyone else (presumably a radio common carrier (RCC) or other applicant) and one-third would be held in reserve to be parceled out later to meet future demands for cellular or some other land-mobile service. ("MHz" and "kHz" mean "megahertz" and "kilohertz" respectively and are used to measure the frequency of radio waves. The standard AM radio dial, for example, falls in the 550 to 1600 kHz range, which is equal to 0.55 to 1.6 MHz—considerably lower than the 800 MHz band set aside for cellular telephone. If you do not understand this technical talk, ignore it. As I explained above, you don't need to know any of it. I just threw it in to impress you.)

A lengthy debate that delayed cellular for even more years revolved around this split, which has been called the "wireline set-aside" or "wireline head start."

The "set-aside" is so designated because half of the usable spectrum is set aside for the wireline carrier in the market. It is also called the "head start" because most markets presumably have only one wireline carrier, while there might be a multitude of non-wireline carriers vying for the non-wireline spectrum.

The Commission announced that it intended to grant the wireline license as quickly as it could and that it would choose the licensee from among the several applications filed with it for the non-wireline frequencies.

Given the nature of the comparative hearing process at the Commission, it became quite obvious that choosing the non-wireline applicant would require a year or more for each city. On the other hand, because the wireline was often able to begin construction sooner, some wireline carriers get on the air and begin serving customers months or years before the non-wireline.

Your cellular telephone will work about as well on either the wireline or nonwireline cellular service. The difference will be local in nature and will include details such as how well the system will be operated with portable phones, how far out the coverage extends into the suburbs and, perhaps most important, how much the service costs.

Even though you may find that the two services are roughly equal in quality, the carriers will work hard to create new features and pricing schedules to attract a larger market share—and perhaps entice your business away from the competing system. You can expect to see advertisements for each of the two competing services that scream and holler "our service is the best." Some differences will, of course, exist between the competing services in any given market, so be sure that you compare both to see which better meets your needs. More on this question in Chapter 7.

Experimental Systems

The United States had two experimental cellular systems on the air prior to the beginning of the operation of commercial service.

Bell Labs and AMPS of Illinois Bell ("AMPS" stands for Advanced Mobile Phone Service, the Bell term for its cellular system) put an experimental system on the air in January 1979, providing cellular service for 2,000 customers chosen more or less at random in the Chicago area. The system functioned perfectly and demonstrated that cellular was a valuable service—that cellular provided a tool of high value to those businesses which were fortunate enough to be selected and put on the system.

The equipment for that system was built by Western Electric, the manufacturing arm of the Bell system.

A second experimental system went on the air in December 1981 in the Washington-Baltimore area, built by Motorola for American Radio-Telephone Services (ARTS), a radio common carrier that serves both cities. (ARTS is now known as American TeleServices Inc.) Motorola is the world's largest manufacturer of two-way radio and mobile telephone equipment.

This Washington-Baltimore system emphasized that portable cellular telephones are viable. It also proved that other manufacturers besides the Bell System were capable of developing and building working cellular equipment.

That system served 300 customers, approximately 200 of whom had Motorola car telephones. The others used the Motorola portable radiotelephones, which we will discuss in Chapter 13.

Meanwhile, as mentioned above, commercial service began in a number of foreign countries including Japan, the Scandinavian countries, Canada and some of the Middle Eastern countries before commercial service began in the United States—ironic because cellular was invented in the States first.

The first cellular system to go on the air in the U.S., the Ameritech Mobile Communications system serving the Chicago area, began operating on October 13, 1983.

Ameritech Mobile is the cellular subsidiary of Ameritech Communications, which is a Regional Bell Operating Company (RBOC)—one of seven RBOCs that emerged at the time of the divestiture of AT&T on Jan. 1, 1984.

The second commercial system in the States went on the air in the Washington-Baltimore area, operated by the non-wireline applicants in those two cities. Five companies filed for the non-wireline system. Realizing that they might be delayed in their effort to get on the air and might lose to the wireline company (Bell Atlantic Mobile Systems), the five non-wirelines combined forces and developed a partnership arrangement to apply for, build and operate a joint system to serve the Washington-Baltimore area. The Washington-Baltimore Cellular Telephone Company, as the partnership is called, began commercial operations on Dec. 16, 1983.

Some observers, in fact, have commented that April 2, 1984, is an even more important day in the history of the development of cellular telephone in the United States than is Oct. 13, 1983, because the April date is the day that Bell Atlantic Mobile Systems put its "Alex" service on the air in Washington, making Washington the first city in the world to have two competing cellular systems.

Washington-area residents were—and continue to be—subject to an advertising blitz comparing the merits of both systems, each claiming to offer more benefits to users.

There has even been some competition on the basis of price. Bell came in at the lower monthly operating fee ($25.00) but the higher per-minute charge (45 cents during peak hours), whereas Cellular One (the name of the service provided by the Washington-Baltimore Cellular Telephone Company) offered its service at the higher price of $35.00 per month but the lower per-minute operating charge of 40 cents during peak hours.

Competition is expected to develop in most of the major markets, and this competition should bring prices down and make cellular service a more affordable bargain for the consumer.

YOUR TIME IS TOO VALUABLE to waste tied up in the traffic jams that plague our major cities every day.

Chapter 4
How Much is Your Time Worth?

Compared to the Value of Your Time, a Car Phone is a Bargain, Not a Burden

You may be under the impression that cellular telephone service is only an expensive telephone for your car. Compared to your home or office phone, cellular is indeed quite a bit more expensive—six or seven times more expensive.

However, to get a truer idea of the cost of cellular service, you should compare its price not to the cost of your home phone but rather with the value of your most precious resource—time.

At the executive and managerial level, time is the most expensive commodity of all. Any tool that can help increase the value of that time is a bargain. For this reason you see companies purchasing their own corporate jets at a cost of 5, 10 or 20 times as much as it would cost to purchase conventional tickets on commercial airlines. The corporate jet helps increase the efficiency of the time of executives, thereby becoming a comparatively inexpensive tool to increase the productivity of those executives.

The same thing applies to cellular telephone. How you calculate the value of your time was the subject of three articles we published in the May, July-August and September 1984 issues of *Personal Communications Magazine.* These articles, "How Much is Your Time Worth? (parts I, II and III), written by time-management specialist Dr. Larry D. Baker, point out that your time is such a valuable commodity that a car phone is a bargain, not a luxury.

This chapter is based on those articles.

This is the longest chapter in the book, but it may well be the most important in helping you decide to purchase your car phone.

How to Determine the Value of Your Time to Your Company

How valuable would a cellular telephone be to your company? Would it be an expensive luxury or a time-saving necessity?

Many businessmen regard the mobile telephone as a luxury, not a necessity.

Remember, not so long ago "the company telephone"—one per building—was a novelty. Everyone knew where it was located, but not everyone was permitted to use it. Placing or receiving a call was a rare experience.

Fortunately, we are accepting new technologies more rapidly today than we did in the past. Yet the spread of the cellular telephone will be subjected to the same uncertainties and "foot-dragging" as were automobiles, electric power, automated assembly lines, central air-conditioning, computers and other similar advances in technology.

A cellular mobile telephone today costs $1400 to $2500. Users can also expect to pay $200 or $300 for installation, about $25 to $40 a month for "dial tone" and about 40 cents a minute to use the phone during prime business hours.

At those prices the cellular phone may sound like a luxury, but before rendering judgment, you should contrast its cost with its benefits.

Any purchase that improves operations, sales, services and, ultimately, profits is by definition not a luxury. Here are some points to consider when you make your decision.

Employees who rarely leave their office and who work only during regular daytime hours do not need cellular telephones. Most employees fit this description.

Obvious candidates for car phones are those who spend time traveling in cars or trucks—those who frequently visit "off-site" locations where regular wireline phones are scarce. Less obvious candidates are those who *should* be visiting customers or field locations but who stay in their offices in order to be "available" when needed. Employees who could expand their workday by using their commuting time to return calls might also profitably use cellular phones.

Not Everyone Needs One

However, merely because an employee spends work time away from a telephone is not sufficient justification to provide him with a cellular phone. His communication requirements and the value of his work time must be considered.

A mobile telephone can reduce the time that an employee would lose while he or she is away from a wireline telephone. Office time that would have been used to place and to receive calls can now be used for other important activities. The value of time saved away from, and in, the office must also be determined.

Managers, salespeople and service representatives are examples of those who spend a large percentage of their work time in vehicles.

Studies indicate that managers spend more time communicating than they spend on any other activity. More than 90% of a manager's time is spent communicating, and 65% of this time he or she is conversing, either face-to-face or by telephone. The remaining 35% of the time he or she is dealing with written communications. Managers who spend considerable time traveling need to analyze thoroughly the communications demands of their jobs.

The use of mobile telephones would permit these managers to communicate with bosses, secretaries, subordinates and co-workers while the managers are away from their offices. Executives would not have to search for pay telephones to relay urgent information. They could turn travel time into productive time by placing and returning calls while on the move. They would be instantly accessible to their secretaries and others to receive critical messages.

Travel time can be quiet time. You can use it to think, plan and organize. Managers who spend more than an hour or two a day away from their office telephones can fulfill many of their communication requirements with a mobile telephone and still have adequate quiet time.

Establishing guidelines and effective call screening by secretaries, message centers and others can do as much to reduce unnecessary telephone interruptions on a mobile telephone as it does for an office telephone. The per-minute charge for telephone use is an added incentive to eliminate unnecessary phone calls.

Excellent for Sales Personnel

Sales experts report that it is not unusual for salespeople to spend only 20% to 30% of their day in direct contact with customers. A mobile telephone might be just the answer for salespeople with significant travel commitments.

Salespeople are likely to spend a great deal of time traveling from one customer to another, often using pay phones and hotel phones to qualify customers, make or confirm appointments and indicate sales reports to the home office.

The mobile telephone might also allow a sale to be made to customer "C" while traveling from customer "A" to customer "B."

A mobile telephone would allow the salesman to do two things at once—drive to an appointment and place such calls at the same time. The less time salespeople have to spend in telephone booths and hotel rooms, the more time they have to spend with customers.

Other candidates for mobile telephones are construction engineers and architects who spend considerable time traveling between construction sites. These professionals can handle telephone calls while traveling rather than spending time making calls after arriving at the sites or returning to their offices. General contractors could benefit from mobile telephones in much the same way as any person having supervisory responsibility for work crews at several locations.

Nationwide and international business that crosses several time zones creates unique communication problems. Many managers and professionals must work extra-long days when they communicate with people in different time zones. A car equipped with a mobile telephone could add from one to two hours a day of productive time to the workday by allowing them to place calls while driving to or from work to companies in different time zones where the work day is still underway.

Mobile telephones could also lead to improvements in health care services and speed their delivery. A doctor driving to a hospital to handle an emergency could be in continuous telephone contact with emergency-room personnel, giving directions for emergency treatment and patient preparation.

Veterinarians could give emergency care instructions to farmers, ranchers and pet lovers. A member of an ambulance crew might be in continuous telephone contact with a victim or those assisting the victim. In emergencies such as these, receiving assurance and being told what to do as well as what not to do can be critical to the happy outcome of the crisis.

Scores of examples could be given to identify candidates for mobile telephones. It is important to gather the data necessary to make the right decision. The question to answer is, "Will a mobile telephone make a sufficient contribution to the quantity and quality of work or increase profits sufficiently to justify its cost and operating expense?"

Analyzing Your Mobile Needs

A communications-requirement analysis and a time-value analysis can help you decide whether you or any member of your staff needs a cellular phone.

Consider the communication log in Figure 4-A. You may need to

---

A DAILY LOG will help to determine if an employee needs a cellular phone. The traveling employee's log will indicate how many calls he receives while he is away from the office and the relative value of those calls to the company. A companion log kept by someone in the office will indicate the number and nature of calls received by the traveling employee while he is out of the office.

Figure 4-A

Name:
Date:

COMMUNICATIONS LOG—MOBILE TELEPHONE NEEDS ANALYSIS

PERSON(S)	HOW *T/V	INITIATED Me/Other	TRAVELING Yes/No	TIME Occur.	HOW Long	PURPOSE—TOPICS	WILL A MOBILE TELEPHONE HELP? How/Why

*T - Telephone
V - Visit

modify it to make it compatible with the position you wish to analyze.

To obtain all the information necessary, an employee who is traveling should keep a log. Someone in the office should record another log for him throughout the day. The second log will indicate which calls and visits he missed while he was away from the office.

Combining the information from both logs will answer the following questions:

- Who are most conversations with—the boss, subordinates, customers, suppliers...?
- Was the conversation a result of a telephone call or a visit?
- Who initiated the contact?
- How many calls were made away from the office? What was the purpose of these calls?
- How many calls or visits occurred while the person was unavailable because he was traveling?

Based on these logs, analyze your answers to the following questions:

- Is there a pattern to the calls and visits? When are they usually made? How long do they usually take?
- How does the pattern relate to travel requirements and office time?
- How much time was spent at the office placing, returning and receiving phone calls that could have been handled with a mobile telephone?
- What important functions and activities were not completed today because office time was used to deal with communications that could have been handled with a mobile telephone?
- How much of others' time was wasted while they waited for information or directions?
- How much time was spent in a hotel room or at a pay telephone making calls that could have been handled with a mobile telephone? What was the purpose of these calls?
- What opportunities have been missed or delayed?
- Who normally calls while the traveler is out? What was the purpose of each call?
- On the average, how much travel time each day could be turned into productive time through the use of a car phone?

Typically, all the data necessary to decide whether to purchase a mobile telephone could be collected in two weeks or less. It is important that the data be applicable to the person for whom the mobile telephone is being considered.

How to Calculate the Value of Your Time

The value of that person's time must then be considered. More than 70% of the managers, professionals and supervisors who attended Time Management Center's time-management programs underestimate the value of their time—from 200% to 500% below its real worth. Ignorance of time value alone could result in the rejection of a mobile telephone. But guessing is not necessary. A highly accurate approach can be used to compute time value. The form titled "What Is Your Time Worth?" (Figure 4-B) illustrates a method of computing value per minute of one's time.

Those responsible for employee benefits can readily identify the percentage of these costs spent by the company for benefits. A percentage for administrative overhead and, if desired, for any additional operating costs related to a given position can be determined. Perquisites must also be included.

Costs of a secretary are included in the analysis because the basic purpose of a secretary is to expand the capacity of the secretary's boss to fulfill the boss's responsibilities. Often, the costs of administrative assistants also are included in the analysis. Many executives have several layers of employees below them, but the analysis includes only the personal office staff.

The break-even costs of an employee must be increased by an amount equal to the company's net profit percentage. This figure represents the contribution to profit that must be made by the employee.

The number of days, hours and minutes a person works can be determined with little difficulty.

When the figures have been assembled, it is a simple process to compute the value of a person's time per minute using cost and profit figures and the minutes worked. The value-per-minute covers all costs associated with employment plus a contribution to profit.

Improving "personal productivity" and "achieving excellence" are in fashion as management concepts of the 1980s. Using your time productively is the key means of achieving these objectives, according to most authorities.

Cellular radio telephony itself has emerged as a unique new tool to increase your productivity by turning "lost" time spent in your car into time that can be used profitably.

Every minute of the working day, an employee costs you money. Each employee must create value through his or her work—minute by minute—to cover these costs. If the value the employee creates is

what is your time worth?

Your Cost:	1. Annual salary and bonus	$ 40,000
	2. Fringe benefits (20% to 40% of salary)	14,000
	3. Overhead (50% to 100% of salary plus fringe benefits)	54,000
	4. Extras and perks	1,500
	5. Your Total Cost	$ 109,500
Secretary's Cost:	6. Annual salary and bonus (Use appropriate percentage for shared secretary)	$ 15,000
Note: Your secretary's time is added in because the only justification for having one is to enable you to produce more than you could without a secretary.	7. Fringe benefits (20% to 40% of salary)	5,250
	8. Overhead (50% to 100% of salary plus fringe benefits)	20,250
	9. Extras and perks	300
	10. Secretary's Total Cost	$ 40,800
	11. Your Time Value at Breakeven (Item 5 plus item 10)	$ 150,300
Your Profit Contribution:	12. Your time value at breakeven multiplied by the company's expected net profit percentage	$ 6,012
	13. Double item 12 to allow for taxes.	12,024
	14. Your Total Value to Company (Item 11 plus Item 13)	$ 162,324

Value per Time Unit:

15. Total days in year — 365
16. Subtract: weekends 104
    vacations 20
    holidays 12
    sick days 10
    personal 5
    total 151
17. Total days worked per year (Item 15 minus Item 16) — 214

18. Value per day (Item 14 divided by item 17) — $ 758.52
19. Value per hour (Item 18 divided by total hours worked per day) — $ 94.82
20. Value per minute (Item 19 divided by 60) — $ 1.58

©1982, Time Management Center, 3855 Lucas and Hart, Suite 223, St. Louis, MO 63121, 314/385-1230; reproduced by permission.

Figure 4-B

A SAMPLE CASE showing a simple method of calculating the corporate value of the time of a $40,000-per-year employee is indicated on this chart.

not at least equal to the costs he or she incurs, the result is a loss for the company.

Because more than 90% of a manager's time is spent communicating (face-to-face or by telephone), each minute he is away from his telephone reduces the opportunity to fulfill the communication requirements of the job.

The form "What's Your Time Worth?" (Figure 4-B) illustrates a simple method of computing the value of your time and of your employees' time. The analysis is divided into four parts: (1) your costs, (2) your secretary's cost, (3) your profit contribution, and (4) the value per unit of your time.

The illustration that follows is based on reasonable assumptions for a manager in the United States. This analysis does not require use of any peculiar figures. All of the data can be derived directly from an organization's standard financial information. If you are figuring the value of your own time, use figures that are appropriate for you.

On line 1, write the annual income. In our example, we have a manager earning $40,000 a year, including salary and bonuses. Obviously, this manager must create sufficient value for the organization through his work to cover the salary he is paid.

Less obvious to many is the additional contribution that must be made to cover the items represented by lines 2, 3 and 4. Fringe benefits (line 2) are just as much a part of the cost of an employee as his salary. The $14,000 figure used represents a 35% employer contribution for fringe benefits. Based on 1983 data, the national average is about 37%. Check with your personnel department for the actual percentage or dollar amount spent by your company for its fringe-benefit package.

The overhead cost (line 3) associated with a single employee may be more difficult to compute, but it is possible to determine a percentage based on income and fringe benefits. The costs considered for deriving the percentage figure on line 3 may include the cost of parking space, office space, office supplies, telephone and other equipment.

Ironically, many employees do not realize that the value they create through their work must result in sufficient dollars to pay for lighting, heating, air-conditioning, sewage and water. Offices have to be cleaned. So do hallways, restrooms, elevators, break rooms and the like. It also costs the company to have wastebaskets dumped and trash hauled away. Whether it is purchased or rented, the company's building engenders costs. Periodically the building may need a new roof or painting and routine maintenance.

The percentage in line 3 may be expanded to include some operating costs. Where there is a large capital investment in facilities, equipment and other expensive items that are necessary for the organization to conduct its business, this percentage may far exceed the 100% figure (salary plus fringe benefits) used here. For our illustration, overhead is listed as $54,000.

What are extras and perquisites, and how much do they cost the company? Extras and perquisites (line 4) are items that usually are not available to all employees. Not all companies provide executives with luxury automobiles, parking space and a fill-up when the tank is empty. Not all employees are permitted to use the company yacht.

These special items are typically buried in overhead or operating costs. Our analysis assumes that the cost of these special items should be borne by those to whom they are available. A percentage may be used if desired, or the specific cost actually incurred may be used. We have assumed the modest figure of $1,500 for this illustration.

Line 5 is the total of all costs associated with the manager's employment. In this example, the total is $109,500.

Our analysis includes a secretary, but if the manager also has an administrative assistant or has others assigned specifically to his or her office, the costs of those employees should also be included. To avoid tedious and irrelevant computations, the analysis should stop with the manager's immediate support staff. We assume the secretary is paid an annual salary of $15,000 (line 6).

In lines 7 and 8 the same percentages are used for the secretary as for the manager. Fringe benefits and overhead are $5,250 and $20,250 respectively. Extras and perquisites (line 9) are $300. Line 10 reflects the secretary's total cost—$40,000.

This is an impressive figure. How many secretaries earning $15,000 per year would realize they must create $40,800 of yearly value by their work to pay for their cost to the company? Likewise, not many managers earning $40,000 a year and employing the services of a secretary would realize that they must generate $150,300 per year (line 11) simply to break even. At the break-even point they have covered only their costs of employment—and made no contribution to profit.

Producing enough value to break even might be acceptable for a manager employed by a nonprofit organization such as a foundation or a branch of government. In a typical business, however, a manager must do more than simply carry his or her own weight. He must bring in profit.

Employees usually do not know their company's profit percentage. Those who do know are most likely to know the percentage of after-tax profits. Therefore, our illustration assumes 4% profit after taxes and a 50% tax burden on the company (line 12).

On line 12 appears the additional dollar value that must be created by the manager to make an average contribution to after-tax profit ($6,012). On line 13 this figure is doubled to provide for the 50% tax on company earnings. The entry for line 13—$12,024—is then added to the break-even cost of the employee entered on line 11—$150,300.

Line 14 is the sum of lines 13 and 11—the total value of the manager's time to the company—$162,324. *This is the value the manager must generate through his work to pay for the cost of his employment and to make an average contribution to profit.*

Not all departments and divisions make the same contribution to profit. In some computations, using a departmental or divisional profit percentage might be more appropriate than using the overall company profit percentage.

This section shows how to compute the number of days, hours and minutes worked, and how to place a value on each unit of time. Line 15 shows 365 days a year. In line 16, days not worked are subtracted from the total number of days in the year. It is assumed that the manager does not work on weekends.

The example also assumes that the manager takes advantage of all days off from work. Obviously, many managers do not take all of their vacation days and sick-leave time and some have to work on holidays. Others may not take personal leave, either. On the other hand, some managers may take more time away from work than that shown in our example. The example is based on the average for a manager in the U.S. who has a $40,000 income.

Days away from work are subtracted from 365 to determine the entry for line 17—the number of days worked a year. In this case, the manager works 214 days out of 365.

The result on line 18 represents the value of the manager per workday. This is the total value to the company of $162,324 divided by 214 workdays. The result is $758.52 per day. The value per hour—$94.82—on line 19, is based on an 8-hour workday. In your analysis, use the length of your own average workday.

The employee's value per minute (line 20) is then determined by dividing his value per hour by 60. In our example, the value per minute on line 20 is $1.58.

Thus you can see how, in our example, a manager who makes $40,000 per year is worth *$1.58 per minute* to his or her company. A manager or an executive with an even higher salary would, of course, be worth proportionately more to his or her company.

For many managers, this figure comes as a shock. They had no idea how much money every minute of their workday is worth.

Compared to the value of the manager's time to his company, the 40-cents-per-minute cost of a cellular telephone is almost insignificant.

How Much is Your Time Worth?

What is your own time worth per minute? We've included a blank form so you can do your own analysis (Figure 4-C). If you are like more than 70% of the managers Time Management Center has worked with over the years, you are in for a surprise when you complete this form.

When you consider the value of your time and how many hour you use the telephone, you will have an easier time deciding whether to buy a cellular phone.

You may see that the value of your time is so high that it is economical to turn the unproductive time spent in transit away from your office into productive time by using a cellular radio telephone.

Can you—or your company—afford not to take a closer look at cellular radio? Cellular radio—rather than being an expensive luxury—may be one of the greatest bargains to come along in years.

A Sample Case

In summary, we discover that the person earning $40,000 annually and having a secretary with a $15,000-a-year salary—allowing 35% for benefits, 100% for overhead and about 4% for "perks," working 214 eight-hour days in a year—is worth about $1.58 a *minute* to his company.

Every minute saved by this person by using a cellular telephone gains $1.58 for the company. For example, if he or she lost 50% of the day as a result of being away from the telephone, the value of each minute of the remaining workday would be $2.16. The value that must be created for each minute of non-travel time is reduced for each minute of travel time made productive.

In other words, the cost of a car telephone is a *bargain,* not a burden, to the company.

When travel time is turned into communication time, a person has more office time available to plan and organize, attend meetings, write reports, evaluate performance and fulfill many other responsibilities. This time-value analysis can be used for anyone.

There are many approaches to computing time value. For example, time value for one minute could be computed by dividing the number of minutes worked into total dollars in sales to be attained in a sales

what is your time worth?

Your Cost:	1. Annual salary and bonus	$ _____
	2. Fringe benefits (20% to 40% of salary)	_____
	3. Overhead (50% to 100% of salary plus fringe benefits)	_____
	4. Extras and perks	_____
	5. Your Total Cost	$ _____
Secretary's Cost:	6. Annual salary and bonus (Use appropriate percentage for shared secretary)	$ _____
Note: Your secretary's time is added in because the only justification for having one is to enable you to produce more than you could without a secretary.	7. Fringe benefits (20% to 40% of salary)	_____
	8. Overhead (50% to 100% of salary plus fringe benefits)	
	9. Extras and perks	_____
	10. Secretary's Total Cost	$ _____
	11. Your Time Value at Breakeven (Item 5 plus item 10)	$ _____
Your Profit Contribution:	12. Your time value at breakeven multiplied by the company's expected net profit percentage	$ _____
	13. Double item 12 to allow for taxes.	_____
	14. Your Total Value to Company (Item 11 plus item 13)	$ _____

Value per Time Unit:	15. Total days in year		365
	16. Subtract: weekends	104	
	vacations	_____	
	holidays	_____	
	sick days	_____	
	personal	_____	
	total		_____
	17. Total days worked per year (Item 15 minus item 16)		_____
	18. Value per day (Item 14 divided by item 17)		$ _____
	19. Value per hour (Item 18 divided by total hours worked per day)		$ _____
	20. Value per minute (Item 19 divided by 60)		$ _____

•Time Management Center, 3855 Lucas and Hart, Suite 223, St. Louis, MO 63121, 314/385-1230; reproduced by permission.

Figure 4-C

HOW MUCH is your time worth? Calculate the answer yourself, using this chart and following the procedures outlined in this chapter. (Source: *The Time Management Workbook* by Merrill E. Douglass, published by the Time Management Center; reprinted by permission. For information, contact the Center at 3855 Lucas and Hart, Suite 223, St. Louis, MO 63121, 314/385-1230.)

manager's territory. Value of goods produced per minute could be used for production managers. Per-minute value could be computed for monthly, quarterly or annual sales objectives for salespeople.

You can use other approaches to calculate the value of time, but any computation that indicates the value of a person's work is likely to lead to a sound decision about buying a mobile telephone.

Examples of How Valuable a Car Phone Can Be

The coming of the cellular telephone has brought us truly into the age of practical mobile communications. With the average cost of a car phone hovering around $2,000-$2,500, however, is the cellular telephone an expensive luxury that is beyond the reach of most business users?

Now that you have calculated the value of your own time, you will know that the answer to this question is a resounding "no!" As you now know, the car telephone is no longer just an expensive toy for the rich. It is one of the most cost-effective business tools that has come along in years. Even though its price may seem unusually high when compared to the cost of a standard desk telephone, compared to the value of your time the car phone is a bargain.

In this section, we will review a number of case studies of users who have added phones to their cars to show just how a car phone can increase your productivity and your profit and more than pay for itself.

For some purchasers, a cellular phone will be a luxury. For others it will be an absolute necessity. Let us look at two business users and several personal users and see how the owners of the cellular phones arrived at their decisions to purchase phones for their cars. These are true stories.

Case 1: A District Manager

Bill J. is a district manager for a retail credit company. He is responsible for 11 branches, each run by a manager. Bill's responsibilities include producing overall profit and ensuring that each branch is profitable.

He draws a monthly salary and can earn a significant year-end bonus based on his productivity. The success of his operations, his future advancements and his annual bonus depend directly on his ability to train good branch managers.

Bill is accountable for maintaining policies and procedures at each branch. These include proper processing of credit applications, maintaining a favorable company image and adhering to all personnel policies.

He makes nearly all advertising and business-development decisions. He hires and trains all branch managers and helps them select and train assistant managers. Hiring and training demand a great deal of his time. Bill also decides what equipment and facilities to purchase or lease.

Credit approval also falls to him. This important duty often requires an immediate response in order to maintain a competitive edge. Branch managers have limits on the credit they can approve—a limit that is based on their longevity and experience. Bill also has a limit on the credit that he can approve. Beyond the limit, applications are referred to the home office.

Branch offices in his district are widely scattered. The most distant branches are about 110 miles apart. Bill tries to visit each branch at least two or three times a week. A new branch or a branch with a new manager is likely to need visits more often.

Every Friday, Bill draws up the schedule to visit his branch offices for the following week.

He needs to accomplish several things on each visit. He reviews operations with the managers, reviews and confirms procedures, and provides training for the managers and their assistants.

He also tries to help the managers with local advertising and with business development. He routinely reviews customers' credit applications with the branch staff. Often an application will sit on a branch manager's desk for Bill's next visit—in order to get his approval or attention before it is sent to the home office.

When people want to buy a new automobile or a new appliance, they're anxious to take possession of the product as soon as possible. To the buyer as well as to the salesperson, no news on a credit application is bad news. Most salespeople are on commission. If they don't get quick decisions, they soon take their credit business to another finance company. Responding to these urgent demands for credit approval keeps Bill and his managers jumping.

Bill begins work each morning about 7:30 a.m. Usually he drives to a distant branch for his first visit and works his way toward home throughout the day. Each day he starts out with good intentions and knows what he wants to accomplish. His days rarely run smoothly, however.

Bill readily admits that important activities not having to do with credit approval—such as training and reviewing operations—often get put on the back burner.

A dozen telephone messages may be waiting for him when he arrives at a branch office. Five or six messages may have been from other branch managers, one or two from retailers, a couple from the home office and some from miscellaneous sources. In addition, he often has one or two phone calls of his own to make.

Bill may intend to spend only an hour or so at a particular branch, but the phone calls alone may take an hour. No time is left for training, reviewing operations or anything else. "I often look at a few applications at a branch and then head for the next one," he says regretfully.

"Recently my day became an entire mess. Before the end of the day, I had to drop one branch from my schedule and cancel appointments with two retailers.

"The bad part was all the unfinished work still sitting on my desk the next morning. I had to take care of it immediately."

On one occasion, for example, Bill had to drive to the farthest branch to solve a problem. He spent two-and-a-half hours in the car. As he was leaving, he asked the branch manager to call three other branches and cancel his visits. Each of them had credit applications to discuss with him or applications needing his approval. At one of the locations he also had to cancel his appointment with a contractor and an interior decorator who were remodeling the branch. Bill got the credit applications approved, but he was on the telephone until 8:30 p.m.

If he had more time to train managers and their assistants, he might not need to discuss applications with them as often.

Bill had to find a way to handle the communication demands of his job and at the same time free himself to take care of responsibilities that he was neglecting. He was just about at his wit's end when he saw a newspaper ad for cellular telephones. It occurred to him that a cellular phone might allow him to turn his travel time into productive time.

Before he discussed the idea with his boss, he wanted to make sure that a cellular telephone would be a good investment. The value of the phone had to be greater than its cost of approximately $2,000 and the usage charges of about 40 cents per minute.

Bill used the communications log (Figure 4-A) to analyze his job.

Written Record Tells the Story

For 2 weeks Bill kept a log of the way he spent his time and the phone calls he made and received. He found the following:

1. He spent from an hour and 50 minutes to 4 hours and 35 minutes each day in his car. He estimated that 85% to 90% of his travel time could be put to productive use on his car phone.
2. He visited an average of three branches a day.
3. At the branches:
   a. He would find two to eight phone calls about credit applications waiting for him. Each call would last about 6 minutes each, for a total of 108 minutes a day.
   b. He made two calls at each branch. The calls averaged 5 minutes each, for a total of 30 minutes.
   c. He made calls to or received calls from the home office four times a day. Each averaged 6 minutes, for a total of 24 minutes.
   d. His calls to retailers, business associates, equipment and facility people and others averaged three per day at about 7 minutes each for a total of 21 minutes.
   e. His total telephone time per day averaged 183 minutes—or about 3 hours.

Losing Business

4. Of the five credit applications withdrawn by retailers during the two-week analysis, four were attributed to branch managers not reaching him quickly enough for approval. In those four cases, he believed a mobile telephone would have saved the business.
5. If he had a cellular phone, he thought, on some days he would not have to leave home as early in the morning, and he frequently could leave for home earlier at the end of the day.
6. Using the techniques outlined in this chapter, he found his time is worth $1.64 a minute. Thus, his 3 hours of average daily travel time cost his company $295.20.

    He realized that the $295.20, which represents one third of the value of his workday as unproductive time, must be added to twice that amount to determine the time value of the six productive hours of his 9-hour workday. In other words, he must be one and one half times as productive during his productive hours to make up for the unproductive time lost while traveling in his car. The value of each minute of his productive hours then becomes $2.43.
7. After they had seen his figures, his boss and the home office encouraged him to purchase the cellular mobile telephone as a pilot project.

Time and Money Saved

Bill collected the following information during the first three weeks he had his cellular mobile telephone:

1. Bill now handled about 80% of his telephone calls while traveling—calls that he would have returned when he got to the branches.
2. On the average he freed up approximately 45 to 50 minutes that he could use to spend at each branch. He now uses this time to review operations, train subordinates and develop more and better business.
3. In the first two weeks, he handled four urgent credit approvals while on the road using his cellular phone—transactions that will earn a considerable profit for his company. Any of the four could have been lost if they had not been promptly approved. One of those transactions, for example, involved a $24,000 loan on a luxury car. The interest income to the company on that one transaction will more than pay for the $2,000 that was spent on the cellular telephone. Two or three such transactions per year would more than pay the monthly cost of operating the cellular telephone.
4. Seven retailers had commented on the speed with which credit applications for their customers had been handled. This improved service and the resulting goodwill suggests that turning Bill's nonproductive time into productive time will ultimately result in an increase in business.

5. Bill is no longer losing $1.64 in time value each minute that he travels. The time he saves by using the phone has freed him to engage in other managerial responsibilities and has made the investment in the mobile telephone worthwhile.
6. Bill's branch managers are happier. Their customers are more satisfied and the managers have more time to spend with Bill when he visits.

Case 2: Laughed out of the Meeting

A few months ago Bob M. attended a district sales meeting for his company. During a business-building discussion, he suggested that placing mobile telephones in the cars of its salespeople could increase the company's annual sales. "I was laughed out of the meeting," he said.

Recently, Bob attended a time management seminar that I was presenting. Mobile telephones came up in the discussion as a way of turning travel time into productive time. Bob responded to that idea immediately, and I sent him a copy of the May 1984 issue of *Personal Communications Magazine*. I suggested he follow the recommendations for collecting information about his telephone communication activities and his travel requirements.

Bob is in industrial sales. His products are highly technical and often are developed to meet customers' specific needs. Still, competition is strong.

Bob's total sales each year are nearly $20,000,000. He makes about 50 sales a year, and each sale averages about $40,000.

His territory is in the industrial heartland of the U.S. He spends 3 out of every 4 weeks calling on local customers. The fourth week he makes an "upstate" sales trip in which he covers nearly 1,000 miles.

He spends at least one day and frequently part of a second day each week in the office. Every week he spends 1 to 2 hours in a sales meeting. He spends most of the remaining time he is in his office on the telephone.

A More Productive Way

He believes much of his office time could be spent more productively—visiting present customers, qualifying potential customers and visiting prospects.

Bob places an average of eight telephone calls each day when he is traveling and visiting customers. Friday is usually his day to be in the office, and on Fridays he places twice as many calls as he does other days of the week.

Whether he is in or out of the office, he usually receives about five telephone calls each day. These are calls that he learns about throughout the day when he calls back to the office for messages.

When he travels, he makes his calls from pay phones. He finds them wherever he can—in restaurants, in service stations, at street corners and in hotels.

Bob frequently spends as much as 30 minutes at a pay telephone—and sometimes up to an hour. Often, noise around him makes it difficult to communicate. If he places a call and the other party is busy, he either has to wait at the pay phone until the party is available or drive on and make an additional stop later.

"The minimum time I spend getting off the highway and trying to find a pay telephone and place a call is about 10 minutes," he says. On a very busy day, Bob may spend 2 hours at pay telephones.

Time Going to Waste

When he is on his 1,000-mile trip each month, hours pass while he drives on interstate highways—wasted hours.

Bob believes that a mobile telephone could save him time in a number of productive ways:

1. His customers, boss, secretary and the technical people at his company could call him on his mobile telephone.
2. He would not have to use pay telephones. He would not have to spend the time looking for them or put up with disturbing noises.
3. He could qualify prospects while he travels rather than use his office time and pay telephones for that purpose.
4. Because of the nature of the products he sells, he must qualify all requests for samples. He usually does this by telephone. With a cellular phone, he could do it as he traveled.
5. He could call and discuss competition and competitive strategies with other salespeople in the company.
6. He could follow up with customers on items requiring action.

Bob believes that with a cellular phone he would be able to reduce his office time each week to about half a day. The rest of the time he could visit customers and prospects. He thinks also that the time saved each day by not having to use pay telephones would allow him to schedule at least one additional appointment.

Topping the Competition

Bob hopes his competitors don't catch on to the idea of the mobile telephone too quickly.

Only one additional $40,000 sale—his average order—would be sufficient to pay for a $2,000 cellular mobile telephone and to pay service charges for many months of use.

At the time this article is being written, Bob had yet to acquire his mobile telephone. His boss, however, became enthusiastic about the potential benefits of a mobile telephone after Bob had showed him his communication log findings for only one week. The two men are preparing a proposal to have Bob conduct a pilot project with a mobile telephone. The proposal will be submitted to their national sales manager.

In the seminar Bob attended, he learned about time-value analysis. To more accurately represent what Bob and his boss believe are costs specifically associated with his job, the overhead cost section of his analysis was expanded to include some operating costs that will make his time value per minute analysis more accurate.

Bob estimates the ultimate figure will be somewhere between $1.60 and $1.80 a minute. Turning part of Bob's office time and roadside telephone time into customer time could prove to be a profitable decision.

Personal Use—Mobile Telephones

Competition in the cellular telephone industry has already begun to drive prices down. Lower costs of equipment and air time will make the car telephone attractive for personal use.

Many users will buy cellular phones because of their "gimmick" or "conspicuous consumption" attraction. Others will buy the phones based upon more practical needs and interests.

Case 3: On Call

Two plant security supervisors had to take turns being on 24-hour call. When a supervisor is on call, he has to be available by telephone and within 30 minutes of the plant. That meant these supervisors stayed home when they weren't working. If they went somewhere else within 30 minutes of the plant, they had to be sure a telephone would be available, and they had to stay in that location.

Then one of the supervisors bought a mobile telephone. Now he enjoys a great deal of freedom. Anytime day or night he is free to travel anywhere within 30 minutes of the plant. He lets the plant security center know when he is traveling and then notifies the security center when he arrives.

The other supervisor is now considering buying a mobile telephone.

Case 4: Keeping In Touch

A housewife whose elderly mother lives with her found a new degree of freedom when she bought a car telephone. Her mother was not critically ill but had limited mobility. Both felt more secure when the mother could call the daughter anytime the daughter was away from the house. The mother always knew the telephone number of the daughter's destination, and she knew the daughter could be reached while in route.

Case 5: Commuting Parents

A husband and wife who commute to work together were concerned about leaving their 10- and 12-year-old children at home before school and having them at home alone after school. The mobile telephone helped. The parents knew that the children could get in touch with them at all times. In the evening, a call to the house from the car could confirm that everything was well with the children.

That call gave their mother an opportunity to tell them how to start dinner.

Case 6: Extended Weekends

A man who owns his own company bought a mobile telephone because he and his wife enjoyed spending long weekends at their cottage. The cottage did not have a telephone.

Now, he can be in touch with his office two or three times a day on the Mondays and Fridays he spends at the cottage. His consulting business requires that he personally travel to his clients' offices, which are scattered over a fairly large area. He now handles a lot of business through his secretary as he travels.

Case 7: Just To Be Accessible

A couple likes to take evening and weekend drives. They use call-forwarding on their home telephone to transfer calls to their cellular mobile unit when they are away from their house.

They admit that the mobile telephone has no real dollar and cents value to them. They simply wish to be accessible by telephone when away from home.

Many other uses will be found for mobile telephones. None of them will be as exciting as discovering your own new ways to make life more profitable and more pleasant. It takes only a little imagination to discover how you can obtain a lot of satisfaction from your cellular telephone.

(Author's note: Dr. Baker's articles end here.)

Another Way to Look at the Value of Your Time

How does the cost of a car telephone compare to the cost of using pay telephones?

After all, local calls may be relatively inexpensive with a pay phone—from 10 cents to 25 cents in most areas for the first 3 to 5 minutes.

Compared to the amount you have to slip into the coin slot, a cellular phone call at 40 cents a minute is expensive.

Depending upon how you value your time, the cost of a cellular telephone in your car can actually be a bargain—not an overpriced luxury—especially if your job requires you to travel frequently by car.

Assume, for example, that you make about four calls per day from pay phones. We will also assume that each call lasts about five minutes. Four calls a day times five days in a week equals about one-and-one-half hours that you will spend each week on a pay phone.

Each of those calls will require you to look for a pay phone before you can make the call. To find the phone will take you 5 or 10 extra minutes, plus the 5 minutes of driving time that you will lose because you had to stop driving to place the call.

In other words, a single 5-minute phone call placed from a pay phone will cost you 10 to 15 minutes of driving time.

Thus, for the one-and-a-half hours a week you spend on the pay phone, you will actually have invested 3 to 4 (or more) hours of time that could be put to more productive use.

If you had made those calls from your car phone, the total cost to you per minute would be 40 cents times 90 minutes—a total of $36.00. Adding in the prorated cost of the service plus the cost of the phone, you find that it will cost you about $50 to $60 per week to use your car phone.

By contrast, even if your time is worth only $30 an hour—which is a very low estimate—that 3 to 4 hours you had to spend looking for and using pay phones would have cost you $90 to $120.

Thus you can see that the cost of a car phone can be considerably less than the amount of time you would have to spend looking for and using pay phones.

Cellular Telephone: An Expensive Luxury or a Bargain?

If you compare the cost of your cellular call to the price of a long distance phone, the cellular phone call comes off as a bargain.

A call from Washington to Los Angeles, for example, would cost you $3.15 for three minutes if dialed from a pay phone using your telephone company credit card. On the other hand, dialed from your car, the three-minute call would cost only about $2.81.

Even greater savings can be attained by using your cellular phone to call across what are toll boundaries within the landline system but which are within the local calling area of the cellular service.

For example, a cellular call from Havre de Gras, Md., which is about 25 miles north of Baltimore, to Lorton, Va., about 20 miles south of Washington, is a local call when placed over Cellular One, the non-wireline system serving the Baltimore/Washington area.

The cost of placing a cellular call between Havre de Grace and Lorton is 40 cents per minute during peak hours (7 a.m. to 7 p.m.) If the same call were dialed directly over the landline telephone network at the daytime rates, the charge for a direct-dial call would be 54 cents (plus tax) for the first minute plus 35 cents for each additional minute.

The cost of a 3-minute call between the two locations would be $1.24 (plus tax) for the landline call and $1.20 for the cellular call. If your call went just 5 seconds longer, the landline carrier would charge you for an additional minute of time, or a total of $1.59 plus tax. Cellular One, on the other hand, bills in 6-second (or one-tenth of a minute) increments and would charge $1.24 for a 3-minute and 5-second call.

Although it is true that few calls would actually end up costing more over the landline network, these figures do indicate that a cellular call need not be any more expensive than a landline call.

The moral of the story is that if you live on the fringes of a cellular-calling area, it may actually be cheaper for you to place your calls to the other side of the calling area via your cellular phone than over the landline circuits.

Less Expensive than Opening and Equipping an Office

Although it is true that a cellular phone—at least for the present—costs considerably more to own and use than a conventional wireline telephone, installing a phone in your car may, in fact, save you far more money than the cost of the phone itself.

This savings can be particularly opportune for the independent businessman who operates a business primarily from his home. As his business begins to grow, he may contemplate the idea of opening an office and hiring a secretary to answer his calls for him. The cost of an office can be $500, $600 or $700—or more—per month plus the cost of the secretary or receptionist.

An alternative might be to install a car phone. When not out on the road, he could answer his own calls at home. When he goes on the road he can put his home phone on "call forward" and have all calls diverted directly to his car, or he could use a telephone answering machine to take his calls and access the answering machine remotely from either his car phone or any other phone. In this way he would save himself the cost of an office.

Compared to the $1,000 or so that the office and secretary would cost, a car phone at $100 or $200 a month could be a considerably cheaper option.

Conclusion: A Cellular Phone Can be a Bargain

When making your decision on whether or not you can afford a car phone, do not compare the cost of the cellular phone with the cost of the phone sitting on your desk at work or on the wall at home. Compare it to the value of your time.

You will then see that the cellular phone, rather than being an expensive toy, is one of the greatest bargains to come along in years.

NO MATTER WHAT YOUR BUSINESS, if you spend time on the road, a cellular phone can increase your productivity. The phones shown here were made by E.F. Johnson.

Chapter 5
How Others Are Using Their Cellular Telephones

Once upon a time—back in those dark ages known as the 1950s—a car owner who was tired of owning automobiles that guzzled gas and gave him three-and-a-half miles per gallon, purchased a small foreign car.

After driving the car for a couple of months, he was disappointed. "I thought this little bug was supposed to give me 40 miles per gallon," he said to one of his friends who also owned a bug and had played a role in convincing him to buy one of his own.

"How is it that you are able to tell me you get such high mileage?" he asked.

"Well," his friend replied, "if you really want to get the high mileage that bug owners brag about, you can do like the rest of us do."

"What's that?"

"Lie about it."

*   *   *

The first cellular telephones that go into people's cars are, I predict, going to cause a great many people to become truth stretchers.

Truth stretching is not exactly like lying. It's not exactly like white lying, either. It is, in fact, a fun exercise for people who are otherwise too lazy to get involved in muscle stretching (a la Jane Fonda).

The truth stretcher who gets a cellular phone in his car will tell tales in an effort to turn you against car phones. He will complain that—contrary to advertisements—the phone is not always as clear as the phone in his office. He will scare you with tales about all those near misses he has had with Mack trucks while talking on his car phone. He will moan and groan about his huge telephone bills.

Do not be alarmed by his stories. What he's trying to do is to scare you away from obtaining your own phone. He knows that the car phone is one of the most valuable business tools that has come along in years—and he isn't about to let his competitors get ahold of one any sooner than necessary if he can possibly prevent it.

(For the record, cellular phones do occasionally have a little noise in them that you won't hear on a landline phone, but it is slight and does not interfere with your conversation. Also, the car phone need not be a safety hazard at all. We'll discuss that question in Chapter 10. And as far as the "huge telephone bills" are concerned, you know the answer to that question after reading Chapter 4.)

In order to give you some idea of what a car telephone can do for you in your business, it might help to examine the experience of other car phone owners to see how they have benefited from this service.

Contractor Doubled His Business

A building contractor in Chicago who was an early customer on the AMPS experimental system told me that after he installed a phone in his car, he doubled his business within a year.

Before installing the car phone, he had found it necessary to spend long periods of time in his office returning and placing phone calls. Once he put the phone in his car, he was able to make and receive all of his calls while in transit between building sites. Thus, he was able to put more effort into overseeing the construction of seven different sites.

Lawyer Increases His Income

A lawyer who had a 40-minute commute between his home and his office in downtown Chicago said that he was able to add two hours of billable time to his day by returning his calls while commuting and while driving between the various court houses. He could now spend that time, which previously had been dead, on the phone with his clients.

As Abraham Lincoln said, "A lawyer's time and his advice are his only stock in trade." A cellular phone may not be able to increase a lawyer's ability to give advice, but it certainly does increase his ability to spend more time with his clients.

A $2,000 Per Month Phone Bill

A real estate agent in Chicago reported that her average monthly car telephone bill is more than $2,000. This figure may seem high—but she finds that it is a real bargain. Her cellular phone has enabled her to more than double her sales—and keeps her in the multi-million-dollar sales bracket.

For example, it's not unusual for her to take a client out to examine one piece of property, drive past another property that looks promising and see a "for sale" sign in front of it. She merely pulls over to the side of the road, dials the phone number on the sign and makes an appointment on the spot, saving time that might otherwise be spent looking for a pay phone.

She also reports that the presence of a phone in her car gives her a certain status in the eyes of her clients and makes them more willing to buy from her than they would be to buy from another agent who does not have a car phone.

She tells the story of trying to close a million-dollar real-estate deal by phone while racing to the airport to catch a plane. She made the deal *and* the plane. The commission from that one transaction more than paid her phone bill for the entire year.

Other Examples

A fast-food-chain executive estimated that his company could save a million gallons of gasoline a year if his peers also had car phones. His savings in driving time translated into savings in fuel—and dollars.

A trucker told AMPS (Bell's experimental cellular program) that he was able to save hundreds of hours of waiting time each year by calling ahead from his truck before arriving at a dock for pickup. He calls a company ahead of time and asks them to have his order ready at the loading dock when he arrives. As a result, he just drives up, loads his truck and leaves. He no longer has to spend 10, 20 or 30 minutes or more waiting for his shipment to be brought to the dock.

A glass salesman reported that his phone had enabled him to double his sales volume. Because he spends an average of 10 hours a week in his car, he is able to use that time to place orders, sell (or presell) and follow up on orders. He is now able to spend more time with the customers and to visit them more frequently.

A doctor who relaxes by playing tennis and golf finds that the combination of a beeper and a car phone increases his opportunities to catch an extra game or an extra set because he knows that if an emergency arises he can be reached immediately. Thus he now enjoys his leisure time more because he knows he doesn't have to worry that he'll be unavailable if an emergency arises.

Time is Money

"Time is money," the old cliche goes. A product or device that can help you increase the value of your time is a money saver as well as a time saver.

For example, depending on the business he is in, it can cost a salesman an average of $200 just to make one sales call. By calling ahead, that salesman can save time and perhaps avoid the embarrassment of arriving late if he gets lost. He can use his car phone to ask for directions and perhaps even be talked to the door of the place he is going. When he arrives, he will be relaxed rather than harried.

Once the sales call is completed, the salesman is in his best frame of mind to act on the information that he has absorbed during that call. If he can pick up his phone as soon as he gets in his car and call in his order or follow up with additional information, the item can be on its way to delivery faster. The result will be a more productive salesman and a happier client.

On the other hand, if the salesman has to wait until he gets back to his hotel room or the office or until he can find a pay phone, he could lose hours or days. Also, the salesman's memory of the client's particular needs will diminish as the amount of time between the sales call and the order placement increases.

The car phone is not just for the chairman of the board and the president of the company. It can be a useful tool in the hands and in the cars of lower-level managers and salesmen and in the vehicles of lower-level employees such as truck drivers.

If the status a car phone carries makes you reluctant to give it to your lower-level people, look on it as a bonus or a perquisite—a way of telling a trusted employee that you value his service. It can be much less expensive than a raise and can immensely increase the person's self-esteem and his prestige in the eyes of his family and his friends.

Your car phone can put extra hours into your day and extra dollars in your pocket.

Those are facts—and you won't have to lie about them.

Chapter 6
Why You Should Purchase Your Car Phone NOW

If you need a car phone now, don't wait "until the prices drop."

CAR TROUBLE? Your car phone can be a real lifesaver.

As more and more businesses add telephones to the cars of their executives and salesmen, the competitive edge that a car phone gives you today will diminish. Consequently, it is important for you to purchase and begin using your car phone now so you can retain that competitive edge as long as possible. Don't wait for lower prices.

You may, for example, have heard reports that cellular phones will be much less expensive to purchase "next year." After all, pocket calculators that cost hundreds of dollars when they first came out are now available for just a few dollars. Therefore, you may be tempted to wait until the price of a cellular phone gets down to a more "affordable" level.

I myself have participated in spreading this rumor of how car phones will cost less "within a year." The lead story in the June 15, 1984, edition of *USA Today* quoted me as saying, "Prices [of car phones] should be down to around $1,000 within a year."

That particular quote caused more of a ruckus than I expected. Our phone rang for several days from cellular equipment vendors and service providers. The standard comment I heard from them was, "Here we are trying to sell these phones at $2,500—and you're telling people to wait a year and not buy them yet! What are you trying to do? Put us out of business?"

The fact is that I was not making a prediction when I made that statement to the reporter. I was merely doing my job as a journalist by reporting on what I had been told by at least two manufacturers who have already announced that they intended to bring out cellular phones at a cost of about $1,000 within that time frame—perhaps even sooner.

Prices of electronics products have been dropping in recent years. Should you wait and buy your cellular phone after the price comes down?

No!

To wait until later to buy your car phone might be a major mistake. If you need a car phone today to aid you in your business, you should buy that phone today—regardless of the price. Do not think of the car phone as an expensive luxury. Think of it as a time-saving and time-management necessity, as explained in Chapter 4.

Even if you do wait and the cost of the phones drops below $1000, what are you going to do then? Are you going to buy one?

Probably not. Why? Because you will open the newspaper and read predictions that "cellular phones will be priced at $750 within a year." So you will decide to wait another year.

More car phones; prices head down

By Mark Lewyn
USA TODAY

Competition is heating up across the USA in the $1.5 billion cellular telephone market.

New York is about to become the latest market; Los Angeles introduced the high-tech car phone system Wednesday, and Philadelphia, Minneapolis, Houston, Dallas, Phoenix and Atlanta are next.

More than 12,000 customers use cellular phones in eight cities. By 1990, projections are 2½ million of us could be using car phones.

Most cellular phones now cost about $2,500 to buy. But, "prices should be down to around $1,000 within a year," says Stuart Crump Jr., publisher of *Cellular Radio News*.

Analysts expect the cost of using a car phone, however, to remain high. Average monthly bill now: $100 to $200.

"I don't expect the price of using one to change for a couple of years," says Clifford Higgerson of L.F. Rothchild, Unterberg, Towbin.

Later this year, a competing system expects approval: Personal Radio Communication Service, which allows users to make calls from their cars at a fraction the cellular cost.

The PRCS system routes calls from your car directly to your home telephone. The cellular system hands your calls to phone company antennae which dot many cities.

Reprinted with permission of USA Today, June 15, 1984 edition.

CAR PHONES can benefit many users—from the company president to the field worker. This photo shows a user on the Nordic Mobile Telephone System, which serves the four Scandinavian countries. The phone is made by Ericsson Communications.

And after that year passes, you will see predictions that "cellular phones will be priced at less than $500 by next year."

The point is that no matter what price you pay for your car phone today, there will be a less expensive one available tomorrow. If you wait for a less expensive one, you will *never* buy the car phone—and consequently you will never be able to benefit from owning and using it.

Furthermore, remember that the price of the mobile unit is only one part—perhaps even the least expensive part—of the cost of your cellular service. Monthly service and usage charges still constitute the bulk of your cellular phone costs (see Chapter 7).

In short, you should select your car phone based on a careful assessment of your mobile communication needs—and not the cost of the mobile unit itself.

[For the record, my "prediction" of the $1000 cellphone came "true" more quickly than anyone expected. In November 1984, the legendary "Madman" Earl Muntz began selling one brand of cellular phone in his Van Nuys, Calif., consumer electronics showroom at the retail price of $995 (plus $100 for installation). Retailers in several markets quickly matched—and even beat—that price. As the year passed, Muntz continued to cut retail prices. At the time this book went to press, he was even offering one model for $495.]

[Cynics may point out that my "prediction" involved *list* prices and also implied that *most* cellphones would be priced at the $1000 level—

which is not, in fact, what happened. Nonetheless, when you are in the prognostication business, you are allowed to twist the facts around and "edit" your predictions later so that you come out sounding like you know more than you actually do.]

Benefit from Your Car Phone Today

Perhaps you are primarily interested in having a car phone for "status" rather than "business" reasons. Even if you intend to purchase a cellular phone primarily for its status value—as many people are expected to do—you should make that purchase *immediately* rather than waiting until the price drops.

Why? Because once the price drops, sales of car phones will increase at a brisk pace and you may lose the status edge of "being the first."

A car telephone is not just a gimmick, a toy or a status symbol—even though it may be viewed this way by many people. It is an important business tool that can increase your productivity and your performance on the job and give you one or two extra hours a day that otherwise might be lost while you are commuting or stuck in traffic.

Time is money, and time is an irreplaceable resource.

In his best-seller *Megatrends,* John Naisbitt writes about "information float." Information float is that period between the time you request information and the time that information is supplied to you.

Naisbitt gives the example of a letter. If you write a letter to someone, it takes three or four days for it to get to the person. The letter may sit on his desk for a day or two while he decides what to do with it. Presumably he will get back to you within three or four days. The entire period of the information float for a letter could be a week to 10 days.

In other words, information float is the period of time in which your request remains in limbo. A telephone call can cut information float to a matter of minutes. You pick up the phone, dial, ask for your information and receive it almost instantly—assuming that the other person is there when you make the call.

If an idea occurs to you while you are driving, the information float is however long it takes you to get to a phone and act on that idea. If you have your phone with you at all times and the people you deal with have a phone with them at all times, you can reduce this float period to practically nothing. Information (as well as time) is money in the modern business world.

In fact, in many businesses, information is the *only* commodity they have to sell. Therefore, the cellular phone can increase their productivity—and their profits—immensely.

THE CELLULAR TELEPHONE is not as expensive as you might think. The phone is by Mitsubishi, marketed in a briefcase model by MAGNUM-ROAMX.

Chapter 7
How Much Does It Cost?

Calculating Your Cellular Phone Bill

Whenever I talk to a group that is new to the cellular concept, the most frequently asked question is, "How much does it cost?"

My standard answer is "about $150 to $200 a month." That price shocks most people and even turns many of them off.

As we have already seen, the price you pay for cellular service—in the $150 to $200 per month range for most users—is largely irrelevant because those dollars can pay you back many times over in increased productivity and output.

You should think of cellular as an inexpensive, productivity-enhancing time-management tool rather than as an expensive luxury. Therefore, price alone should be a minor factor in deciding whether or not you need a car telephone.

Nonetheless, most potential users still want to know how much cellular service will cost. The prices in this chapter are those that were in effect at the time that this chapter was written. I believe that they will decrease over the next few years. How far and how quickly is anyone's guess. By the time you get around to purchasing your car phone, you may be pleasantly surprised to discover that it won't cost you as much as you thought it would.

The Three Parts of Your Cellular Telephone Bill

Your cellular phone bill, like Gaul of Caesar's time, is divided into three parts.

1. **Equipment Cost.** The first part of the bill—and initially the most expensive part—is the cost of the car telephone itself. At the time of this writing, car phones were available at prices ranging from about $1,500 to $2,600, although some dealers were quoting retail prices of $1000 or less (plus antenna and installation). When the first phones went into service in Chicago, in October 1983, the average user paid

about $2,500 to $2,900 for his phone.

When competition came to Washington, D.C., a couple of dealers cut prices to less than $2,000 within 3 months after service began. By early 1985, one retailer had dropped the price on its bottom-of-the-line model to $995 plus installation.

Competition is hot and heavy in the cellular market. As discussed in the previous chapter, several vendors have already announced that they plan to market cellular phones at prices below $1,000 in 1985 or 1986—but don't wait to buy your phone. If you need a car phone now, buy it now. You may save $300 by waiting six months—and lose a $10,000 or a $100,000 sale because you weren't near a phone when a customer called to place an order.

If you have a large fleet of cars or trucks to equip with cellular phones, you can expect to receive a large discount from your service supplier. For example, one phone may cost you $2,400, but if you buy 20 you may be able to get them at a cost of $1,800—or less. The latter figure would probably also include installation, which usually adds another $200 to your price.

Likewise, if you have a fleet of a hundred trucks or cars in an area, you should be able to order your service at bulk rate—the same rate charged to resellers. Hertz, Budget and National rent-a-cars have ordered bulk service in several cities for their rental fleets, as have a number of taxi and limousine fleets.

2. **Monthly Service Charge.** One part of your bill will reflect a flat monthly fee. You could think of this as a charge for "dial tone" (even though, technically speaking, you will not hear a dial tone when you dial your cellular phone).

This charge does not vary no matter how much—or how little—you use your phone each month.

This fee can range from about $10 per month up to about $45 or $55 depending on the market you are in. Also, individual carriers offer a number of different plans in most markets, and each plan may offer a different monthly service charge.

For example, carrier "X" may offer "Plan A" for the "regular" user, with a monthly service charge of $40 and a per-minute charge of 40 cents peak, 25 cents off-peak. "Plan B" might be designed for "infrequent" users and offer a monthly service charge of only $10 but a higher per-minute charge of 75 cents for every minute the phone is used—plus possibly a much lower charge for off-peak usage, perhaps 15 cents a minute. This type of plan is designed to make phone ownership more attractive to non-business users.

Believe it or not, carriers in some markets are actually offering plans that charge you *nothing* for your monthly service charge; you pay only a per-minute charge for actual air time you use. Ask your carrier, and pick the plan that best meets your needs.

3. **Per-Minute Usage Charges.** You will also be billed for each minute of airtime that you use the system. These charges vary widely, but for the present time you can expect to pay during peak periods of the day (usually 7 a.m. to 7 p.m.) from about 25 cents to about 45 cents for each minute of time that you use the phone. This 25-to-45-cent charge is in addition to any long-distance charges. Most systems will offer you a discount on the charge during "off-peak" periods.

Billing practices also vary widely from carrier to carrier. In one metropolitan area, for example, at the time of this writing, the non-wireline carrier charges $35 per month for service plus 40 cents per minute for service during the peak hours. The wireline carrier charges $25 per month for basic service plus 45 cents per minute for peak-hour service.

Another difference between the two carriers was mentioned in Chapter 5. The wireline bills for usage in whole-minute increments, whereas the non-wireline bills in tenth-of-a-minute segments. In other words, if your call lasts for 2 minutes and 10 seconds, the wireline will bill you for a full 3 minutes of airtime. The non-wireline would bill to the nearest tenth of a minute—or 2 minutes and 12 seconds.

Furthermore, the wireline has structured its tariff so that you are charged even if the number you have dialed is busy or if there is no answer. Non-wireline customers are not billed for no-answer and busy-signal calls.

Also, under the wireline tariff, you will be billed for one minute of service every time someone calls *you*, even if you are out of the car and do not answer the phone.

This particular pricing policy could prove to be quite expensive. If you received just three calls a day that you were unable to answer because you were out of your car, and if you placed three calls a day that received a busy signal or went unanswered, you could find yourself paying $80 to $100 extra per month on your bill *for service that you do not receive!*

This approach may not seem to make sense to you as a user, but from the telephone company's point of view, it does have a certain logic behind it. The theory is that the service *is* available to you and you are using expensive air time even if you do not actually answer the phone. Therefore if you do not pay for that air time, those costs will have to be spread out among all the other customers.

Despite this reasoning, many observers believe that the public will not stand for this method of tariffing. In other words, if there are two carriers in the market and one charges you for busy signals and no answers and the other does not, the no-charge company will most likely attract the largest number of customers.

These differences between the two carriers in this particular market (which I have deliberately not identified) were in effect at the time this chapter was written. Competitive pressures may well change these differences by the time this book goes to press. Be sure to check with your carrier or carriers for the latest information.

My comments here should not be construed as a recommendation in favor of one carrier or against another in any given market. The fact that one carrier charges less than or in a different way from the other may not necessarily mean that one carrier is better than another. You will have to make your own decisions on which carrier is the better one for you based on the pricing policies in effect in your area at the time you purchase your phone as well as your own needs.

As more and more cities have two competing systems, the carriers will compete with each other to see who can come up with better, more creative tariffing practices. The long-term winner in this business will be you, the user.

Where to Buy Your Car Phone

Unlike traditional telephone service, you will be able to select your car phone service from at least two underlying carriers. Each will have a number of agents, resellers and dealers selling their service to the public.

Your options are as follows:

A. The underlying carrier. As explained in the Chapter 6, one and possibly two cellular carriers will serve each market. One will be the local telephone company or a subsidiary of the local telephone company. The second could be anybody else, such as Cellular One, MCI, a Metromedia subsidiary, a local radio common carrier (RCC) or a partnership including several of these.

It might even be a fruit company. United Fruit, for example, applied for licenses in several cities. United Fruit has a subsidiary, TRT, which is an international communications carrier. You may find yourself talking on a bright yellow Chiquita phone. (If you think the country is going bananas over cellular, you may be right.)

You will probably not buy service directly from the local carrier. In most cities the carrier will be selling through its agents and resellers.

B. Agents. Carriers will select their own agents to market their service directly to the public. Agents are directly controlled and regulated

A SAMPLE CELLULAR BILL covers a variety of charges.

by the carriers.

If you sign up for service through a carrier's agent, you become a customer not of the agent but of the carrier itself and you receive your bill from the carrier each month. If the agent goes out of business, the carrier will continue to take care of you. Agents are carefully screened by the carriers before they are selected.

You will be able to tell if the seller is an agent because it will state that point specifically in its advertising. For example, "Authorized Cellular One agent" and "Authorized Alex agent" appear frequently in ads in Washington.

C. **Resellers.** Resellers are primarily self-selected. They are entrepreneurial individuals or companies who see cellular telephone as an opportunity to provide a personalized service to the public.

Resellers purchase blocks of numbers and air time at a discount from the underlying carrier and resell to the public at a markup. Resellers depend upon your satisfaction because they know that they must follow up after you have purchased your phone from them. They bill you directly. Resellers often have their own agents who operate in a manner similar to that of carriers' agents.

Therefore, you may find the independent reseller to be an even better company from which to purchase your phone because it depends on your good will and your repeat business. The reseller also hopes that you recommend cellular services to your friends and thus bring him additional business. (Word-of-mouth referrals are one of the primary sources of new sales in the cellular business.)

The unfortunate side of the reselling business is that—in some markets—resellers have had an extremely difficult time making a profit.

In Chicago, for example, four companies started out as resellers, and at least one company has gone out of business. The underlying carrier in that market, Ameritech Mobile Communications, initially sold phone numbers and airtime to the resellers for only 2% less than it was charging its own customers. The resellers attempted to make their profit on the sale of the cellular equipment and installation, but as those prices became competitive, the resellers' profit margins were driven too low. Ameritech has now improved margins it offers resellers so that they are closer to the national average.

In other markets, however, the carriers are providing larger margins of profit, which means that resale in those cities is a viable business.

It may be difficult to tell from an ad whether a company is a reseller or a dealer.

D. Dealers. Dealers are generally retail stores that sell only mobile equipment and do not sell the service. For example, you could go to a Sears, select a cellular phone from the shelf and pay for it. Sears would then install it for you in your car. The Sears store would then make provision for you to get your phone number, but the store itself would probably not be an agent or a reseller. Instead, it would make its own arrangements with one of the agents or resellers to provide you with your telephone number.

The dealer makes his profit on the sale of the telephone plus the installation. Occasionally he may also repair a phone, but he will not as a general rule make a profit on the air time that you use.

On the other hand, an agent or a reseller might be willing to give you a lower price on the phone if you plan to it use it a great deal. Agents and resellers make a small profit on every minute of air time.

Be sure to check several agents, resellers and dealers before making your purchase.

One thing you should *not* do is select your phone on the basis of price alone. You may find a discount house that will sell you a cellular phone for a very low price but which has no facilities to maintain it. The discounter may tell you to take it to an "authorized dealer." The authorized dealer will be less than enthusiastic about repairing your phone if he did not actually make a profit on the initial sale. He may charge you more than he would have charged one of his own customers. As they say on Wall Street, investigate before you invest.

Chapter 8
How to Select Your Car Phone

A Smorgasborg of Features to Choose From

STANDARD AND OPTIONAL features such as hands-free operation, lock and horn alert, are available on many phones, such as this AudioTel phone by Audiovox.

Such a wide variety of cellular phones are available on the market today that you have a multitude of features to choose from. Shop around before buying.

As of this writing, Motorola had just brought out its bare bones "POTS" ("plain old telephone service") cellular telephone, which offers basically nothing but a phone and dialing buttons. It works perfectly well and would be fine for the individual who uses his car phone only occasionally and does not wish to invest a lot of money in it. This unit, however, would not be adequate for the executive who wishes to use his phone constantly because it does not include such things as memory dialing or an LED readout that lets you know what phone number you have dialed.

Some of the features that you may want to look for when you select your phone include:

Push-button dialing. The earliest automatic car telephones had rotary dials, just like the dials on an old-fashioned desk telephone. Those phones, however, have largely been replaced with push-button Touchtone® type dialing models. All cellular phones have push buttons for dialing.

Dialed-number display. The number you dial is displayed on an LED readout across the top of the phone. Some readouts have only seven numbers on them. Some have 10. Some can display as many as 20 or more digits at one time. This longer display can be useful if you make a lot of long-distance calls over an alternative carrier such as MCI or Sprint. For most purposes, the 10-digit display is adequate. The bare-bones phones such as Motorola's 2000 have no display at all. A display is not necessary for those who use the phones infrequently.

On-hook dialing. Sometimes called "pre-origination" dialing, on-hook dialing allows you to dial your number while the phone is still in its cradle. The number is displayed on the screen. When you finish dialing the number, you can check to make sure that it is correct. When you push the "send" button, the number is automatically transmitted over the air.

The advantage of pre-origination dialing is that you do not waste valuable air time while dialing. Some systems will charge you for every second after you push the "send" button—even if you don't get an answer. Therefore you want to keep that time to a minimum. All cellular phones have this feature.

Repertory dialing, sometimes called "memory dialing." I consider this an absolutely essential feature. This feature allows you to place from 9 to 99 or more numbers in various storage areas of the telephone. You pre-program the phone so that, for example, hitting the "recall" button followed by a "4" might bring up your office number, "5" might bring up your home number and so forth.

The advantage of this feature is that it allows you to dial seven or 10 (or more) digits at the touch of one or two buttons on the keypad. This can be an important feature to help you dial safely while making calls as you drive.

The disadvantage of this system is that it requires you to remember which number or numbers are the abbreviated form of the longer numbers you intend to dial. I have found it useful to make the one- or two-digit recall numbers equal to the first digit or two of the telephone number that I intend to call.

In other words, if my home number is 227-XXXX, to recall my home number I would hit "Recall-22." If my office is 255-XXXX, I would hit "Recall-25," and it would bring up my office number. This technique works well as long as the numbers you are calling are in different local telephone exchanges.

Another method is to dial the numbers using the *letters* on the keypad. For example, if you have a car phone and your partner also has a phone in his car, you could store his phone number under 22 because the "2" button on the phone also contains the letters "C" and "A", which are the first two letters of the word *car*. You would simply dial "Recall-CA"—which is the same as dialing "Recall-22"—and it would bring up his car-phone number. You could put the office number under 63—or "OF" on the dial, the first two letters of "office." "Home" would go under "46," which is equal to "HO," the first two letters in "home." If you call John Jones frequently, you could store his number under "JJ"—or "55"—in your memory dialer.

This repertory dialing feature can be quite useful because 80% or 90% of the calls you make will probably go to the same three or four numbers. In fact, because most of the calls will go to the same numbers, you may not need 99 memory locations. Ten or 20 may be sufficient.

Dial-in-handset verses dial-in-base. Some models put the Touchtone® dialing pad in the base of the phone. Others put the pad on the handset (usually on the back of it.) I prefer the dial in the handset because it allows someone in the back seat to dial the phone while holding the handset. It is also a bit more convenient to bring the dial closer to me while attempting to drive and place a call at the same time. Most phones are equipped with the dial-in-handset. You make your own decision as to which you prefer.

The dial-in-base option may be the better choice for you if you intend to make frequent use of the...

Electronic scratch pad. You are driving along at 55 mph, talking to your secretary. She tells you to call Mr. Smith at 555-4567 immediately after hanging up with her. How do you write down a phone number and drive at the same time?

No problem—if you have an electronic scratch pad. As she dictates the phone number to you, you simply punch it into the keypad on your phone. As soon as you complete the call, you hit "recall." The number 555-4567 pops up on the display screen. You hit "send," and the phone automatically dials your call to Mr. Smith.

Privacy muting. In the old days, when you were a kid, you may recall how your mom used to put her hand over the speaker portion of the phone when she shouted for you to come to the phone. She was practicing a simple form of privacy muting—a method still widely used today on landline phones.

This method is not too practical for use in a car where you need to keep at least one hand on the steering wheel. Hence, cellular manufacturers have instituted privacy-muting buttons. When you push the mute button down, it cuts out the microphone on the handset. This way you can talk privately to someone else in the car. The person on the other end of the line will not hear what you are saying.

This feature would probably be unnecessary if you drive alone most of the time and seldom have passengers in the car while on business.

Last-number recall. You are driving down the highway at 50 mph and you dial an unusual number—one that you call very seldom. You get a busy signal. Because you rarely call the number, it is not likely to be in one of your repertory-dialing memory locations. When you want to dial the number again, you can use the feature called "last-number recall."

Last-number recall will bring back from the phone's memory the number that was last called and place it in position so that you can dial it again with the touch of a button or two. This is an extremely useful function.

A note of warning, however. In some systems you may be charged for a full minute of air time every time you push the "send" button even if you get a busy signal. Check this potential problem out. If such is the case, you may not wish to try the same number over and over again every 20 seconds. Instead, you may want to wait 2 or 3 minutes or more between call attempts to decrease the likelihood that you will receive another busy signal—and another bill for 40 cents each time you dial the number.

Horn alert. If you tend to do much of your business outside of but within a short distance of your car, the horn-alert function can be quite useful. For example, if a construction supervisor parks his car at the edge of a construction site and receives a phone call while out of his car, the car horn will honk, alerting him that there is an incoming call. This can be a useful feature for you at the tennis court or at the swimming pool—or even at the office, if your car is parked outside your window. This function can usually be turned off if you are going to be too far from the phone to answer a call.

Electronic and/or manual-key lock. To the unintiated, car phones are extremely attractive toys. When you take your car into, for example, a parking lot and give your key to the carhop to park your car, you might prefer that he not place a 20-minute call to his girlfriend in Sandusky.

A lock can prevent this problem. An electronic lock allows you to lock the phone to prevent unauthorized use by punching a pre-arranged code on the keypad. To use the phone, you have to push the "unlock" sequence of buttons on the phone pad.

A personal note about electronic locks: I almost never use the electronic lock on my portable phone (the Motorola Dyna-TAC 8000X). One afternoon my phone suddenly refused to work. I thought it was broken, so I drove all the way downtown to Motorola to get it fixed. The gentleman who was going to send it out to the repair shop was away from his desk, so I left the phone.

He called me the next day and asked me what my electronic lock sequence was. I told him. He said he would try it.

A few minutes later he called me back and told me that the problem was that it was locked. "So it works now, does it?" I asked. "Sure does," he said. "I'm calling you on it."

The moral of the story is (1) learn *all* of the functions of your phone and how to use them, and (2) if it doesn't work, it's probably not the fault of the phone itself.

A key lock is a useful device as well. It requires you to insert a key into the side of the phone and lock it so that it cannot be used by anyone who does not have the key.

Some locks allow you to lock the phone so that it can be used only for incoming calls. Any locking function can prevent unauthorized use of your phone and unexpected charges on your bill.

Back-lit keypad. The dialing pad on your phone should light up so that you can see what numbers you are dialing at night. Almost all phones have this feature.

Hands-free operation. Speaker phones are quite popular for office use. In the office, a speaker phone allows you to talk on the telephone without holding the phone to your ear.

In a car, hands-free operation can be even more important. No one has yet conducted a thorough study of the safety problems that might arise when attempting to place telephone calls while driving. Simple reasoning will tell you, however, that having one hand tied up holding a phone leaves you with one less hand to operate your car. This problem could be compounded if you were driving a stick-shift car and were required to put the phone down or hold it between your shoulder and ear while negotiating a turn and shifting at the same time. Hands-free operation could be literally a lifesaver in such a situation. I would definitely recommend it.

Sometimes, however, the quality of the sound that you hear over the speaker phone can be diminished if you are attempting to drive with your windows open on a particularly noisy day. Therefore, it is good to have a handset as a backup in case you find the speaker phone unusable.

Another disadvantage of the hands-free phone is that other drivers are likely to give you strange looks as they see you talking to yourself while driving. Ignore them. They're probably just jealous.

Clip-on headset. If you have ever watched an operator at work in front of a switchboard, you know that she (or he) uses a device that clips over her ear and keeps both of her hands free to operate the controls on the switchboard. At least one car phone manufacturer has built a similar headset for its cellular phone—a stretchband that extends over the top of your head holds the unit in place. A small "boom" microphone extends out from the headset and is positioned in front of your mouth. With the headset, you can drive and talk on the phone at the same time and yet leave both hands free to operate the car.

Cordless control head. As of this writing, no cellular car phone is made with a cordless control head. We do, however, expect to see this feature offered in the future. A cordless control head would be similar to a cordless telephone of the type you have at home. It would allow you to use your cellular telephone handset without having it tied by a cord to the base unit in the car. This concept seems particularly appropriate if you intend to use the handset in the front and back seats of the car and don't want to worry about cords that can get tangled up.

It would allow you to walk 100, 200 or even 500 feet away from the car and use your handset—in a sense allowing you to carry your phone with you just as if you had a portable.

So far as we know, no manufacturer has yet announced plans to bring out such a unit. One conventional IMTS car phone manufacturer does make a cordless control head. Two manufacturers told us that they would *not* make such a device because they believe that the future cellular telephone will be a portable phone that will slip into some sort of a cradle in the car while you are driving. It would thus be a self-contained phone that would provide all the functions of a cordless control head without the necessity of building a complete car phone.

The fact is that if you find that you will want to take your phone with you when you leave the car, you should investigate obtaining a portable that can be slip into a separate adapter in the car. (More on portables in Chapter 13.)

Message alert. You are expecting an important call but cannot wait in the car. You need to get out and run into the store for a few moments to pick up something.

What do you do if the phone call comes in while you are not in your car and you will be out of the range of the horn alert? A "message alert" light will help you in this circumstance. If the phone rings while you were out of your car, the message-alert light will come on if the phone is not answered. This light indicates that a call came in while you are out of the car.

At least one cellular phone is available that allows a caller to touch a number into his or her phone's Touch-tone pad; this number will later be displayed to you on the LED read-out on your car phone when you return to your car. Thus you will know who has called you.

Unless you have this display feature, the disadvantage of the message-alert light is that if you receive calls from several people, you will not know which of them called. You may end up calling five or six people to figure out whose call lit up your light. This problem could also be solved with...

Built-in answering machine/dictating machine. Answering machines are quite popular for home and office use with conventional landline telephones. It is only a matter of time before a car-telephone manufacturer realizes that an answering machine can be an important device for the car telephone user as well and builds one into the phone. (Just as this book was going to press, one manufacturer announced that it intends to bring out a small telephone answer machine that will attach to its cellular phone in the car and operate off the car's batteries.)

The same unit could be used as a dictating machine and also a recording device to record your car telephone calls in case you needed to take notes while talking on the phone. At the time of this writing, no such feature was available.

System Features

The features listed above are the type that have to be built into the phone unit itself. Other features are available through your carrier.

Typical system features available now include:

Call-waiting. If you are talking on the phone and someone else tries to reach you, you will hear a click or beep to indicate that you have a second call coming in. Depress the switch hook quickly, and you can talk to the second party. Depress the hook again, and you return to the first call. This feature is essentially the same as having two lines at a fraction of the cost.

Call forwarding. Call forwarding allows you to direct incoming calls to any other phone number. For example, suppose you have driven to a restaurant to have your dinner and are expecting an important call. You could program your phone to forward all calls to the restaurant's telephone. Be sure to inform the manager that you may be getting a call so that when the call comes in the caller won't be told, "Sorry, wrong number."

Three-way calling. While driving down the highway discussing an upcoming meeting with your partner, Harold, you suddenly realize that you need to talk to George at the same time. With three-way calling, it is a simple matter to bring a third party into the conversation and have a three-way conference call.

If possible, the three-way call should be initiated by the landline party rather than by the car party because the person in a car will have to pay double the price for bringing two parties on line. Such arrangements can be done from the car, however, and this can be a useful feature if you need it.

Call waiting, call forwarding and three-way calling are services that you are probably already familiar with because they are available to landline phone customers in many cities today.

OKI ADVANCED COMMUNICATIONS in 1978 delivered the first cellular phones that were used in the initial stage of the Chicago test system, operated by Bell Labs and Illinois Bell. In February 1982 it also became the first manufacturer to receive FCC type acceptance on a cellular mobile phone. The CDL-230 is the company's top-of-the-line model.

GLENAYRE's cellular phone includes a key to lock the trunk unit into the car to help prevent theft.

WALKER TELECOMMUNICATIONS' mobile telephone features a built-in computer interface.

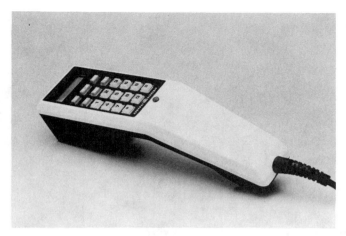

GENERAL ELECTRIC'S cellular phone control head.

PANASONIC's cellphone control head.

THE ALPHA CUSTOM control head by Harris.

THE WEBCOR ELECTRONICS model CH36 cellular mobile telephone.

THE BLAUPUNKT MT 8000 AU cellular telephone.

Chapter 9
Mounting the Phone in Your Car

Convenience and Safety are Prime Concerns

YOUR CAR PHONE should be mounted in a convenient location in the car where it can be easily reached by both the driver and the passengers. In this photo, our company's vice president of sales Don Moore demonstrates the ease of operating the well-mounted phone, manufactured by Western Union.

Make sure that your mobile phone is conveniently mounted in your car. The standard location is on the floor hump between the driver and the front-seat passenger. Be careful, however, that you do not have the phone mounted too low or too close to the floor. You should not have to move your head from your normal driving position to dial the phone. In other words, you should not have to take your eyes off the road to place your call.

A better place to mount it would be on the dashboard just below the car radio, where it will be easily accessible to both the driver and the front-seat passenger.

Some phones allow you to move the handset from one location in the car to another by means of jacks mounted at different locations throughout the car—in much the same way that you can plug your home phone into jacks at different locations throughout the house.

This feature can be useful if you need to make calls from both the front seat and the back seat of the car. You may be able to purchase an extra control head so that a phone can be permanently mounted in both locations. (Both control heads would operate off the same radio unit, which would be mounted in the trunk.)

Be sure that you have an expert perform the installation. It is possible to do your own installation by following the directions. However, if you should, for example, make a poor contact between the antenna and the body of the car, the loss of signal strength can be quite substantial. The poor connection could severely limit the range of your car phone and reduce the quality of your service. Once the phone is mounted, be sure to have it double checked by a qualified engineer.

Antenna. A variety of antenna types are available. In the early trial system in Chicago, Bell Labs mounted two antennas on the trunk of the car to allow for a concept known as "spacial diversity," which increased the quality of the radio signal received at the car.

Further testing proved that two antennas were not necessary—that a single antenna performed quite adequately—so the diversity antenna practice is now an extra-cost option.

That the two-antenna concept was dropped is, in fact, somewhat of a disappointment to many pretenders to cellular status. When I visited the installation center for the Chicago trial system in October 1981, one of the installers told me that the cellular phone had become such a status symbol in the Chicago area that anyone who had two antennas on his trunk was immediately looked up upon as being some sort of big shot.

He told me that a number of people had approached him to ask if

THE PIG-TAIL ANTENNA is standard for cellular car phones. The glass-mounted model by Antenna Specialists, examined here by Jack O'Dowd of Oki Advanced Communications, permits the antenna to be mounted without cutting a hole in the car.

he would install two antennas on the trunk of their car. They didn't care about having the phone itself. They just wanted the antennas so that people would *think* they had a car phone.

If you wish to make sure that everyone knows you have a car phone, you will want to use a special type of antenna—the "pigtail" antenna—which is unique to the cellular car phone. The pigtail antenna has a number of turns in it so it looks like an elongated spring.

If you prefer that your car not advertise the fact that it contains a car phone—which, by the way, is not a bad idea because it will discourage break-ins and thefts—you can also purchase a variety of "disguise antennas." These look like normal car-radio antennas and can, in some cases, provide service for both your car phone and you car's AM/FM radio through the same antenna.

You can also unscrew the antenna and store it under the seat or in the trunk when you park you car in any area where you suspect it might act as a magnet to attract the wrong type of curiosity seekers.

If you do not wish to have a hole drilled in the trunk or roof of your car to mount the antenna, it is also possible to get a new type of antenna that mounts on the glass of the front or rear window without damaging the glass at all. This antenna is "capacitance coupled," which means that the connection is made through the glass without direct wire contact.

Talk to your installer about the various options that are available. My preference would be for the disguise antenna. The disadvantage of the disguise antenna is that it gives a slightly weaker signal pattern than the roof-mounted pigtail antenna, which gives about the best signal pattern available and is definitely recommended if you intend to use your phone in fringe areas far from the system.

IF YOU EXERCISE caution, driving while talking on the phone should be no more hazardous than talking with a passenger in the car.

Chapter 10
How Safe is Your Car Phone?

If You Exercise Caution, Your Car Phone Will Offer You Many Years of Safe Use

A nagging question is heard more and more as the number of phones on the road keep increasing: "Is it safe to drive a car and talk on the telephone at the same time?"

At the time this chapter was written, that question was still being debated. That debate is likely to continue for many years. Some states may, in fact, attempt to resolve the question by passing laws that prohibit the driver of the car from talking on the phone while the vehicle is in motion—a law that is probably unnecessary and would be impossible to enforce.

Two recent studies—one by Ameritech Mobile Communications and one by AT&T—indicate that drivers who have cellular phones in their cars are, in fact, *safer* than the average driver on the road. Speaking on the phone while driving is no more difficult or dangerous than speaking with another person who is seated inside the car with you—if you exercise the proper precautions when phoning.

We touched upon the question of safety when we discussed where to mount the car phone and also in the section on how to use one-number dial and last-number recall.

The smart, reliable driver can easily carry on a conversation with two, three or more people in the car with him. On the other hand, some drivers seem to feel that they are being more polite if they turn their head and direct their vision toward the person they are talking to.

Whether or not it is safe for you to talk on the phone while driving has as much to do with you—the driver—as it has to do with the inherent technology of the car phone. A careful driver will be in no danger when making a car-phone call because he will not let the phone distract his attention from the road. On the other hand, the less-than-careful driver—who takes risks and drives foolishly—will have yet another distraction to contend with as he ambles aimlessly down the road.

All it will take will be a few of these less-than-careful drivers to smash themselves up while talking on their car phones to convince the unthinking bureaucrats that *all* car phone callers are dangerous. The only way to avoid this problem is to make certain that these bureaucrats are among the first to receive car telephones. They wouldn't think of banning something that they themselves use.

Most states already have laws on the books that cover the question of the safe usage of car phones. These are those laws that require you to pay full time and attention to your driving. Thus if you became involved in an accident while using your car phone, you could be charged under one of these laws if the investigating officer could prove that your use of your car phone had prevented you from paying full time and attention to your driving.

This question of the legality of placing a phone call while you are driving has not yet been resolved in the courts, but we may see a test case before too long.

It's difficult to see how a law could be enforced that prohibits you from placing car phone calls while driving. If you read the instructions that come with most car phones, they state that "the driver should not place calls while the car is in motion." They recommend pulling over to the side of the road to place calls or having them placed by a passenger.

This requirement is largely wishful thinking and will not be followed by the vast majority of car phone users. Consequently, you have to take extreme care to keep yourself safe while using your car phone. It will be up to you to make sure that your driving record does not suffer just because you have installed and begun to use a car telephone.

One safety problem that is usually not mentioned is the so-called "curiosity factor." As you drive down the highway placing the call, drivers passing you are likely to become extremely curious to know what you are doing. They may crane their necks to see if they can figure out what it is you have in your hand. Thus they, rather than you, may be the ones who are in most danger.

A hands-free phone solves this rubberneck problem because other drivers would not realize that you are placing a call if they do not see a handset in your hand. They may, of course, wonder why you are talking to yourself, but they are less likely to spend a lot of time staring at you under those circumstances than they are if you are holding a phone in your hand.

Before too long we expect to see hands-free car phones that can be activated entirely by the sound of your voice. When you wish to make a call, you will state a command such as "Phone!" Your voice will turn on the unit. You will hear a dial tone. You will then say, "Dial office" or "Dial home." The unit will automatically dial the number as stored in its memory. For a number that is not stored in memory, you will merely recite the phone number, one digit at a time.

The hands-free, voice-activated phone will represent the ultimate in safety.

Another safety problem that has yet to be resolved is that of taking notes while talking on the phone. A built-in dictation machine with a button that you can push to record your conversation into a standard or mini-cassette tape could go a long way toward solving this problem. Until such a device is available, more than one car phone user has developed a technique of holding a small dictating machine in his hand and dictating his notes right into the machine while talking on the phone.

If the message you need to take down is a phone number, you can use the phone's electronic scratch pad to store that number. If you intend to use this function a lot, a phone with its dial in the base could be quite useful because it would not require you to remove the headset from the speaking position in order to punch in the new number.

Needless to say, taking notes in long hand while driving is practically impossible and should not be attempted. You might wish to take a course in memory to increase your ability to remember what it is you need to take notes on and then write it all down later.

Also, make sure that your phone's dial is illuminated so that you can see the phone after dark.

Finally, make sure that the buttons on your phone are large enough that you can easily see them without fumbling and that the section which displays the number being called is large enough for you to read.

(For a compilation of the latest tips on how to use your car phone safely, see Appendix A in the back of the book.)

A reasonably intelligent driver who is keenly aware of the potential danger inherent in phoning while driving and who makes a special effort to stay alert should have no difficulty in placing calls from his car while driving.

As the bumper sticker says, "Be Alert. The World Needs More Lerts."

AN INEXPENSIVE WAY to test drive a cellular phone is to rent a car containing one. Budget, Hertz and National are among the companies that have equipped some of their rental cars with cellular phones.

Chapter 11
How to Get Free Service— and Other Considerations

If you need a car phone, you could go to your local car dealer, reseller, car stereo shop or wherever and purchase one—and possibly be on the road using your phone within just a few days.

However, you may want to try a different approach.

Leasing Your Car Phone

You do not have to purchase your car phone outright. You can lease it. Within a few months after the introduction of cellular telephone in Washington, D.C., cellular phones could be leased for $75 a month from several suppliers. Those leases, generally for 3 years, allow the lessee to purchase the phone at the end of the lease for a relatively low cost. Typical car-phone prices in 1984 fell into the $2,000 to $2,500 range. As the price of the phones drops into the thousand-dollar range, the cost of leasing your car phone will be proportionally lower. Keep your eye on the ads in your local newspaper.

How to Rent a Cellular Phone

If you would like to try a car phone out for a short period of time to see how much you like it before you buy your own phone, you could rent one for a day or two and try it out.

The easiest way to rent a phone is while on a business trip to a city that has cellular service. Check with the rental car agencies to find out which ones have cars equipped with cellular phones.

Generally you can rent a car with a phone for only about $8.00 or $10.00 more per day than the cost of renting the car itself. You will also have to pay the 50 cents or so per minute for usage.

The Hertz, Budget and National rent-a-car companies have announced their intentions to equip at least part of their rental car fleet with cellular phones in those cities which have cellular service. Other car-rental companies are expected to follow suit. Check directly with the rental firm or your travel agent.

A word of warning. Once you have tried a car phone and used it in your business for a day, you will probably be hooked and want one of your own immediately. Do not resist this temptation.

Buy Your Car Phone at the Time You Buy Your Car

Most car dealers have or will have available the option to purchase your car phone at the time you purchase a new—or even a used—car. Check with the dealer.

The cost of the phone can be added to your monthly car payments. Assuming that a new car will cost you anywhere from $8,000 to $20,000 and the phone will cost an additional $2,000, the extra cost on your monthly car payments need not be more than about $50—or even less if stretched out over a period greater than three years.

RoamerCard
O, Give Me a Home
Roamer Service Bureau

Authorized Cellular Roamer Credit Card
Service Number 910-823-0123-4567

Moe Bill Fone
On-the-Go Co., Inc.
1234 Hexagonal Way
Buffalo, New York

O, Give Me a Home Where the Mobilephones Roam

Roaming is the ability to take your car phone with you and use it away from your home area. For example, if you are based in Chicago and you are driving to Washington, roaming would mean that you could use your phone anywhere that there was a cellular system such as in Indianapolis, Detroit, Toledo, Pittsburgh and Philadelphia.

Eventually, when the entire country is wired for cellular service, you will be able to place and receive telephone calls anywhere in the nation. As of this writing, roaming is not practical in most areas. The different carriers are in the process of negotiating roaming agreements and establishing a national cellular clearinghouse—apparently a more complex problem than it might appear to be at first glance. FCC rules require that roaming be allowed. Thus, the carriers will eventually reach agreements.

Until the entire country is wired for cellular service, you can still put a phone in your car that will allow you to place and receive calls anywhere in the nation. Actually, you will have to place *two* phones in the car to cover the entire country.

Several manufacturers offer an option with their phones that allow both cellular and an IMTS (the older car phone system, which still provides good service in less-populated areas) service using the same control head inside the vehicle. When you are driving in a big city, your phone would operate on the cellular channels. When you drive out into the countryside or into the less-densely populated areas of the country, your phone would switch over to the IMTS frequencies.

If you intend to do a lot of driving in rural, non-urban areas that do not yet have cellular service and you need a phone in your car wherever you go, you should consider purchasing one of the combined cellular/IMTS units. Ask your cellular dealer what model(s) he carries that will cover both services.

Eventually, of course, cellular will be everywhere, and you will need only one phone to go everywhere.

If you travel frequently and take your car or portable cellphone with you, you may want to subscribe to *Cellmaps™: The Roamers Guide to Cellular Systems,* a loose-leaf guide that gives all the details you will need to help you obtain service in cities away from home. For information, contact Cellmaps at P.O. Box 66843, St. Petersburg Beach, FL 33736-9990, or phone 813/345-6150.

...But Not in the Air

Note, however, that you will not be permitted to use your portable phone from a commercial airliner. Why?

Well, first of all, passengers are forbidden to use any radio transmitters by Federal Aviation Administration rules. The theory behind this prohibition is that a passenger-operated transmitter could interfere with the airplane's navigation system and might cause an accident. Whether or not this problem has ever occurred in practice is not known, but the FAA wants to make sure it never does happen.

Second, Federal Communications Commission rules forbid the use of a cellular phone in airplanes. Cellular phones are licensed for use on the ground only, as the FCC will tell you if you ask why. Other frequencies have been set aside for use from air to ground—and they are not cellular frequencies.

Legal aspects aside, your cellular phone may not even work in an airplane anyway. A cellular phone operates through a single cell. If you take that cellular telephone up above the clouds, its signal will be picked up by many cells—and even in many cities—and may confuse the cellular system's central switch.

For example, a cellular portable in a plane flying between Baltimore and Philadelphia at 20,000 feet in the air might turn on cell sites in both the Baltimore and Philadelphia systems, tying up channels in many cells in both systems and possibly providing no service to you.

If you carry your cellular phone with you as you go through the airport passenger-screening equipment, you may be challenged by the security guards. They are most concerned that you not use the phone in the air.

Before going through the security check, lock your phone so that it cannot be used and cannot accidently be turned on while it is knocking around in your luggage. Assure them that you are aware of the rules and that you have already taken precautions to prevent inadvertently transmitting while in the air. They may ask you to remove the battery and leave it with the stewardess or captain. If they do, there isn't much you can do about it.

How the FAA and the airlines ever hope to enforce their prohibition is not known. Perhaps airliners will have to be equipped with an internal all-band receiver that will automatically alert the pilot whenever an on-board transmitter is activated.

Otherwise, they will have to rely on the all-watching eyes of the stewards and stewardesses to keep a watch for phone users. This approach will probably not succeed because passengers are most likely to use their phones during take-offs and landings, and these are the very times that the stewardesses are strapped into their own seats and are unable to roam the cabin looking for illegal talkers.

To make matters worse, the take-off and landing periods are the most dangerous parts of any flight. It is during those times that the risk of interference to sensitive navigation gear is most acute.

The advice from here is:

(1) Do not attempt to use your cellular phone while flying until and unless the FAA changes its rules and regulations, which we doubt will ever happen. (Government agencies being the way that they are, we expect to see the rules strengthened rather than abolished once portable phones come into more common use);

(2) If you must carry your portable with you on the plane, remove the antenna and battery pack before going through security clearance;

(3) Bury your unit deeply in your carry-on bag so it is less likely to attract attention when you send the bag through the x-ray machine. It is better to avoid a confrontation if at all possible; and

(4) If challenged, remain calm and explain nonchalantly that you know the rules and have already taken the proper precautions to avoid any difficulties that might arise. Stress that you carry the phone with

THIS BOOK IS YOURS FREE when you accept this invitation to become a Charter Subscriber to *Personal Communications Report*.

If you enjoyed and profited from *Cellular Telephones: A Layman's Guide*, you will want to subscribe to *Personal Communications Report*, the first telecommunications publication written especially for you as a consumer.

The normal subscription rate to *PCR* is $24 per year, but as a reader of this book you are entitled to the special price of $14.

Fill out and mail today the postpaid card.

For fastest service, *call our toll-free order number today* (1-800-CAR-CALL ext. **89**) and use your credit card (MasterCard, VISA and American Express cards accepted).

RETURN THIS POST-PAID SUBSCRIPTION CARD TODAY

Subscribe now to *Personal Communications Report*

☐ Please enter my charter subscription to *Personal Communications Report* at the special rate of $14 (a discount of $10.00 off the regular $24 rate). If after receiving two first issues I am not convinced that it is worth at least twice what I paid for it, I may request a full refund and keep the issues already received, free.

☐ Check or ☐ money order for $14 enclosed.
☐ Charge my: ☐ MasterCard ☐ Visa ☐ AMEX

Card no. _____
Expiration date _____
Signature _____

Name _____
Company _____
(If delivered to a company address)
Address _____
City _____ State _____ ZIP _____
Phone (including area code) _____

Personal Communications Report

Your guarantee: Read *PCR* newsletter for two months. If you aren't 100% satisfied that it is worth at least *twice* what you paid for it, simply drop us a note. We'll refund every penny. Keep the newsletters with no further obligation.

Fill out and mail today the postpaid card.

For fastest service, *call our toll-free order number today* (1-800-CAR-CALL ext. 89) and use your credit card (MasterCard, VISA and American Express cards accepted).

Call toll-free 1-800-CAR-CALL ext. 89; weekdays, 8:30 a.m. to 5:30 p.m. Eastern time (in Virginia and the Washington, D.C. metropolitan area, please call our local Fairfax number, 703-352-1200)

BUSINESS REPLY MAIL

FIRST CLASS   PERMIT NO. 5627   FAIRFAX, VA

POSTAGE WILL BE PAID BY ADDRESSEE

Personal Communications Report
4005 Williamsburg Ct.
Fairfax, VA 22032

NO POSTAGE
NECESSARY
IF MAILED
IN THE
UNITED STATES

you all the time on many trips and never use it in the air because you know how dangerous that can be.

Do not despair, however. If you absolutely *must* make phone calls while flying, a new in-flight telephone service began in October 1984. Airfone, as it is called, should be available on most major domestic flights within a year or two. The service won't be cheap—$7.50 for the first 3 minutes and $1.25 for each additional minute, but it will provide you with a way to make a quick call ahead for a business or personal reason while still in the air.

How to Get Free Cellular Service

Free cellular service! It is available, but be forewarned that getting free service is extremely difficult. It is, however, well worth the effort if you succeed.

There are several ways to go about it, none of which work all the time and may not work for you at all.

The best way to get free service is to become a "friendly user." Every cellular system goes through a building phase and a testing phase before it goes into full commercial operation.

Before the service can begin operating commercially, the company needs to conduct tests to see how well the equipment is performing. Usually, the company will install phones in the cars of its staff members, but there will probably not be enough staff people to adequately cover the area.

Here is where you come in. Volunteer to let your car serve as a guinea pig to help test the system. Promise to take good notes and report all holes in the system—where coverage is poor, where signals fade, where you can't get coverage and where any other problems develop.

You may or may not be able to convince a carrier to let you have free service during the month or two prior to the start-up date. It definitely helps if you know someone on the inside, particularly in a higher level, decision-making position. Our company, for example, was able to obtain friendly-user service on one of the Washington, D.C., systems for about 2 weeks prior to its actual start-up date. (No, I won't tell you which system.)

A second way to obtain free service is to become a salesman or other representative for the cellular company itself. If you find this service as exciting as I do, you may wish to try this method, particularly if you are in the market for a new job. Naturally, you will have to have a phone in your car in order to sell it to others, and the company you go to work for will provide it for you.

A third way to obtain "free" service is not exactly free—except to you. Let someone else—such as your company—purchase the service for you. Explain to your company in great detail why you absolutely *must* have this service to improve your productivity. After reading this book, you should be an expert on that subject.

A fourth way is to figure out a method of making your car phone a tax deduction. Depending on the nature of your business, the cost of a car phone is probably a legitimate business expense. Talk to your tax lawyer or accountant. Anyone who is intelligent enough to know the value of a car phone is also intelligent enough to figure out a way to get Uncle Sam to pick up most or all of the tab. Write your own tax loophole.

I definitely do not recommend trying any of the "phone phreaking" methods that have enabled certain elements of society to get free service—illegally—on the conventional wireline telephone lines. The phone phreaker views what he is doing as "fun," but the phone company and the law views it as stealing. The phreaker who tries to extend his free-call methods to his car phone may be surprised to discover just how easily he can be caught because radio waves make the user highly visible to law enforcement agencies.

Furthermore, the cellular telephone itself includes features that identify the unit electronically even if the telephone number is changed.

Chapter 12
How to Get the Maximum Status Value out of Your New Car Phone
If You've Got It, Flaunt It

by Sue Easton

HOLLYWOOD STARS such as Elke Sommer were among the first to discover both the status value and the usefulness of cellular phones.

Let's face it, despite the fact that a car phone is a useful and profitable business tool, it is also the ne plus ultra *of status symbols.*

To explore this idea of how to get the maximum status value out of your car phone, we asked humorist Sue Easton to write an article for our magazine on that subject. We liked the article so much that we are reprinting part of it here for your benefit. Her article follows.

* * *

"Every improvement in communications makes the bore more terrible."
—*Frank Moore Colby*

Let's assume that you have, after years of drooling over the idea, finally obtained your first car telephone. Maybe you bought it for yourself. Maybe you were able to convince your boss to buy it for you and pay those outrageous monthly bills that will soon be gracing your mailbox.

No matter how you came to obtain that phone, you are faced with a most difficult problem: What is the proper etiquette that must be followed to break your good news to your friends, business associates and peers? And how can you break that news so that it makes the biggest impact without sounding so pretentious—and boring—that they will no longer speak to you?

After all, you do not want them to view you as a bore and refuse further contact with you—thus totally undercutting the higher niche in the world of status that you were trying to carve out for yourself when you bought your car phone.

Sparing no expense, *Personal Communications Magazine* commissioned a series of in-depth interviews among owners of car telephones to ask them how to tackle this delicate social act. Here are the preliminary results of our survey. (The final survey will be published at some future time and be available at an inflated cost that you will not be able to afford, so don't even ask.)

How to Let Them Know

(1) *List your car telephone number on your business card.* This is a simple enough process and will not be viewed as pretentious. To ensure privacy and save money on your phone bill, add "for emergencies only" after the number.

(2) *Send out pre-printed Rolodex cards with your business name*

and address information on it, including your car telephone number. This is considered standard operating procedure in the business world today when a company changes an important part of its corporate identity such as its name, address or phone number. Be sure to include a cover letter informing the recipient of the reason for the new Rolodex card. Be sure to emphasize that the number is unlisted and *must not be given out to anyone.*

(3) *Mention your car telephone casually in business letters and memos.* ("Give me a call sometime on my car phone. You'd be amazed just how well it works. You'd think you were calling from the office.") These letters also provide an excellent opportunity to dispose of the surplus Rolodex cards you will no doubt have left over after fulfilling point number "2" above.

(4) *Study up on jokes about car telephones and tell them when the moment arises.* (See, especially, the story that ends this section.) It is generally better to make yourself the butt of the joke. Merely having a car phone is heavy enough. There is no need to pile it on higher and deeper. Here are a few comments to get you started:

"I spent 20 minutes the other day looking for a phone booth before I suddenly realized I had a phone sitting right beside me in the car. Can you imagine being so out of it? I'm just not used to the thing yet."

"What I like the *most* [you say sarcastically] is that it costs me *40 cents* every time I get a wrong number—which is what I usually get because it's so hard to dial while driving."

(5) *Let comments about your new phone "slip"* over lunch, at the cocktail hour, on the golf course or at a social gathering—but do it in such a way as to downplay any status value.

As John Brooks, the author of *The New Snobbery in America* and *Telephone,* says, it is de rigueur to "pay tribute to the state of being rich" while making it clear that you are "partly kidding." This is known as "parody display."

Example: "I used to enjoy quiet classical music as I drove home from work, but ever since they gave me that blasted car phone, I'm really learning to hate commuting."

Another example: "The nerve of some people! My insurance agent *doubled* my premium when he found out I was calling him from my car phone. He *said* it was to cover the cost of the phone" (see "corollary" under point number "7" below).

(6) *Leave the receipt for your car telephone clipped to something you are handing to someone.* When he asks, "What's this?", explain, but seem embarrassed by your error.

(7) *Place a call, but don't mention that you are calling from your*

car. During the conversation, cry out in such a way as to make it impossible for the other person *not* to ask what has happened. Your reply should incorporate a rather vivid description of the near-miss you have just witnessed: "Some guy doing 70 just cut me off in my lane and practically ran me off the road." This will get the point across that you must be calling from a car without directly mentioning it.

Corollary: We do not recommend that you use your car phone to call, for example, the local welfare office to ask where your food stamps are. Or if you do, do not mention that you are calling from your car. The old "I'm calling from in front of your door in my car" stratagem may work wonders for vacuum cleaner salesmen, but it is not recommended for communications with certain government agencies or to people such as dentists and insurance salesmen (see point "5" above) who are likely to bill you at a higher rate for their services if they find out you can afford a car phone.

On the other hand, liberal use of the car phone is recommended for calling people such as one's mother-in-law who you would rather avoid calling altogether. Toss in frequent phrases such as, "I'd love to talk longer, but the phone company charges me 40 cents a minute to use this car phone, so I'd better be cutting out. Catch you later." She will be impressed by your frugality.

Also, you could hang up on yourself in the middle of a sentence and claim later that "there must have been something wrong in the circuits; I wasn't able to get through when I tried to call you back."

(8) Finally, if all these subtle approaches aren't your style, you can always *use the direct approach*. Find as many reasons as possible to give people lifts around town. Make sure the most talkative person sits in the front seat.

If he or she doesn't immediately ask you about your car phone, invite him to use it to call his friend—preferably a friend high up in the world of local gossip who will be sure to spread the word about your newest toy. We have never known anyone who, upon placing his first call from a car telephone, did not immediately blab out some comment such as, "Hey, guess where I'm calling from," followed by a better sales pitch than even the best salesman could concoct.

For the blanket effect, start a car pool. (The Department of Energy, if it still exists, will love you for this one.)

Comment casually as often as possible, "If we don't get out of this traffic tie-up, I'll call the office and tell them we'll be late." Be sure to drive through areas where traffic tie-ups are frequent occurrences.

One-Upmanship

Finally, that sad day will come when your effort to keep ahead of your neighbor will slip away because he, too, will have purchased his own car phone.

A good sense of humor can get you through such trying times. Ponder the following story, which we believe was first told by Jackie Gleason in the late 1950s:

"I was riding along in my new chauffeur-driven Cadillac," Gleason said, "when we pulled up at a stoplight right behind Mike Todd, who was also in his chauffeur-driven Cadillac. I picked up my car phone and called Todd in his car.

" 'Hey, Mike,' I said. 'Isn't success wonderful? Here you are, and here I am, riding down the street in chauffeur-driven limousines, talking to each other on our car phones.'

"I watched Mike through his car window, nodding in agreement. 'Fame sure is wonderful,' he said. 'But I'll have to ask you to excuse me for a moment, Jackie. My other phone is ringing.' "

A PROTOTYPE of the type of portable cellular phone you may soon be carrying in your pocket is displayed by Jim Korb, product planning and development manager for Mitsubishi Electric Sales of America. The unit's main antenna is built in. The external antenna pulls out to boost reception in fringe areas.

Chapter 13
The Car Phone vs. the Portable Phone
Which is the Better Choice for You?

Cellular telephone has been promoted primarily as an advanced form of car telephone. In its early stages, that is precisely what cellular telephone is—a superior form of car phone.

Many experts believe, however, that the more promising future for cellular telephone lies not so much in the car phone—no matter how useful such a device is—but rather in the *portable* phone or, as we like to call it, the self-contained "personal communications device" (PCD) which can be carried with you anywhere you go.

If you've never seen a portable phone, the first time you see one you might be tempted to think of it as if it were a standard "cordless" phone.

Cordless phones such as the Freedom Phone or the Extend-a-Phone, which you can buy over-the-counter for less than $100 at Radio Shack, Sears and hundreds of other retail outlets, are small units that allow you to make a telephone call through your home telephone up to 500 or 700 feet from the phone via a low-powered radio link. They have a limited range but do offer a nice service if you wish to take your phone, for example, out to the garden, the backyard pool or the tennis court across the street.

A true cellular portable is not much larger than a cordless phone, but it allows you to take your phone and use it *anywhere* cellular service is vailable.

My Motorola Dyna-T•A•C portable (model 8000X), for example, can be used anywhere in the Washington-Baltimore area. Once nationwide roaming service goes into effect (as discussed in Chapter 11), I will be able to use it in any cellular city that I travel to. This is truly communication on the move.

Most manufacturers have indicated that they intend to market a portable telephone at some point in the development of their cellular models. At the time of this writing, however, only Motorola has produced and marketed a working portable cellular telephone.

I have used a Motorola Dyna-T•A•C portable cellular phone for several months and find it quite useful. The phone does suffer from one major shortcoming that we hope will be resolved in the next couple of years—the battery problem.

Solving the Battery Problem

The primary problem is that there has not been a major breakthough in battery technology for two-way radio operations in 20 to 30 years. The nickel-cadmium battery is still the best rechargeable battery available for use in portable phones. Nickel-cads suffer from several problems such as limited life, limited capacity and weight.

In proportion to the amount of energy a nickel-cad puts out, it is considerably heavier than, for example, a magnesium-alkaline battery. The latter battery, unfortunately, is not rechargeable.

With the Motorola Dyna-T•A•C portable, which weighs about 28 ounces, approximately one fourth of the unit is the battery pack. The unit gives about 8 hours of "on" time with one-half hour of "talk" time between charges. Thus, the portable becomes a useful tool to carry with you between fixed- or car-phone locations, but it is not a type of a phone that you could use constantly all day. The batteries would drain too quickly. (Your bank account would also drain quickly because 30 minutes of use a day would give you a monthly phone bill of about $400. Therefore, having a shorter battery life can, in some cases, be an advantage. It encourages you to keep your conversations *short.*)

Until battery technologies catch up with the micro-electronic revolution, there are several ways to solve this limited battery-life problem. One way is to carry an extra battery pack with you. Another is to use a special type of car adapter such as the one that is available for the Motorola portable. The portable can then be adapted for use in the car and operate off a separate antenna built into the car as well as the car's battery. Also, the portable battery can be removed from the phone and slipped into a charger in the car and be recharged at the same time.

A portable phone gives you true freedom of movement. With a portable phone you can go to the tennis court, to the golf course, to a restaurant, take a bus trip or take the commuter train—and remain in instant touch with the outside world. And, of course, the portable will also work quite well in your car.

Join the Beep Generation

The portable also adds a certain element of the unpredictable to your day because you never quite know when it will ring. It might sound off in the middle of a concert or a movie, where it would disturb other people.

One user in Washington tells of the time he was standing in line at the bank when his portable phone rang. He answered the phone, finished his conversation in a minute or so, and hung up. When he turned around, he noticed that everybody in the bank was looking at him.

As portable phones become more common, this type of problem will disappear. However, a new problem will emerge—the problem of the "beep generation." People will be beeped or paged or phoned everywhere they go.

The Boston Symphony Orchestra already includes a statement in its program that "All electronic paging devices are to be turned off during the concert. Thank you."

You can solve the problem of untimely calls by turning your personal communications device off when you do not wish to be disturbed.

Move the Decimal Point

The chairman of Motorola, Bob Gavin, tells a story of one of his friends who is a "prominent senior partner" of a large banking company in Chicago. This friend saw the Dyna-T•A•C portable and begged Bob to lend him one. Finally Bob could resist no longer. When he heard that his friend was going to be visiting the Washington area, he arranged for him to have the use of a portable for the day. (This story took place in 1982, when only 100 hand-made prototype portables were available and were used for test purposes on the Washington-Baltimore experimental system.)

While walking between the two houses of Congress, the friend's portable rang. It was his secretary informing him that a "large private placement opportunity" had developed for his firm. If he could call the "appropriate decision maker" immediately, the deal would be his.

Without missing a step, he returned the call and closed a multimillion dollar deal using the portable phone—all while walking the streets of Washington. This deal would have gone to some other firm—or may well have not been made—had it not been for that portable cellular phone.

As Gavin explains, "You can tell that story hundreds of times, maybe with a different placement of the decimal point, to get some idea of the impact the cellular telephone will have on the American economy."

This story indicates how valuable a portable phone can be. As Martin Cooper, who wrote the introduction to this book, said: "The portable phone is as great an improvement over the car phone as the car phone is over the wireline telephone." (Cooper is the former research director with Motorola who spearheaded that company's cellular telephone program and who now owns his own company.)

Time alone will prove just how true Cooper's words are.

Do You Need a Portable Phone?

You may not need a portable phone at this time. Whether or not you do will depend upon the type and extent of your mobility.

If the only times you need a cellular phone are while you are driving between your home and your office or between your office and your appointments, a car phone will probably be adequate.

On the other hand, if you live in a large city such as New York and generally avoid traffic problems by not driving—many New Yorkers do not even own a car—then, of course, a car phone will not help

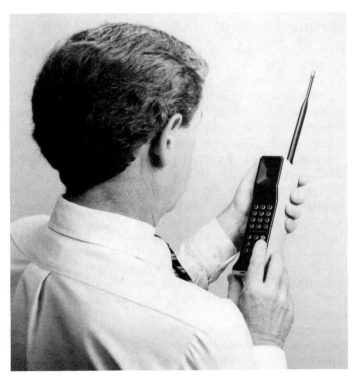

MOTOROLA's DYNA-T•A•C 8000X was the first portable on the market.

you much. You would need to use a portable if you intend to have mobile-communications capability.

You might be tempted to ask, "Why would I want a car phone when I could get a portable phone that will perform both functions for me—as both a car phone and a portable?" The answer is that, as discussed above, the portable does have its limitations at the present time, particularly in its battery life. It also has a much lower transmitting power and will not work in many fringe areas where a car phone will pull in a signal.

Motorola has engineered its system so that a portable at the side of the street will receive and transmit as well as a car phone from the same location. However, when you take that portable into some areas of a building—into the basement or into an elevator, for example—you will find that the "no service indicator" on the phone will light up, informing you that you are outside the range of the cellular system.

The portable has one other disadvantage—price. A portable phone now can cost you twice as much as a car phone. At the time of this writing, Motorola's unit has a list price of $4,000 plus an additional $800 for the car adapter and $60 to $80 for the rapid charger. The rapid charger allows the batteries to be recharged in one hour and is highly recommended if you use your phone a lot. An extra battery pack will cost you another $80 and is also recommended if you intend to talk for long periods of time.

This price discrepancy may diminish within a year or two, however. Several other manufacturers have announced their intentions to market cellular portables. Competition will undoubtedly bring the price down considerably.

Advantages and Disadvantages

The chief advantage of the portable phone is that you can carry it with you and use it almost anywhere you go. Its chief disadvantage is its limited battery life and limited power output.

For the present time, the portable has a part to play in the personal communications revolution, but it is not yet a substitute for the car telephone, and it is certainly not a substitute for the fixed, desk-type telephone. But its day is coming.

The Transportable

Several manufacturers have begun to introduce "transportable" cellular phones. These are essentially car phones that can be removed from the car and used by themselves with an attached battery pack—generally inside some sort of custom-made carrying case. You sling the entire package over your shoulder and carry it with you.

This arrangement can be quite effective if, for example, you make frequent visits to some property you own out in the country where there is no wireline telephone service. You just remove your car phone and carry it into your cabin for use on the weekends. The unit can also be transported easily onto your boat for a spin around the lake.

NEC's Xcell System is a completely self-contained transportable unit. The entire transmitter portion of the phone is contained in a small unit beneath the handset.

The Briefcase Phone

If you like the idea of portability but are turned off by the limited battery life available with the portables available today, you might want to consider purchasing a briefcase cellular telephone. At least a dozen models are available today, and more are being introduced all the time.

The briefcase phone offers most of the advantages of the car phone as well as the chief advantage of the portable phone—its portability—and by-passes most of the disadvantages of each.

The briefcase phone uses a larger battery pack than a portable and gives you a longer total talk time—as much as an hour or more between charges.

The car phone's main drawback is that it is tied to the car. A briefcase phone can be used in a car—but it can also be removed from the car and taken with you wherever you go.

An article in the November 1984 edition of *Personal Communications Magazine* covered the development of the briefcase phone in depth. The following are excerpts from that article, which was written by magazine assistant editor Benn Kobb:

Making the Case for the Briefcase Phone
by Benn Kobb

They say you can't take it with you—but in *this case,* we make an exception.

"As new as it is, the *car* telephone industry is dead," John Fisher explained. "It has been replaced by the briefcase telephone."

Fisher, who is president of Professional Communication Systems of Falls Church, Va., was demonstrating his phone-in-a-briefcase—and predicting the demise of the cellular phone in its role as a car telephone.

"It [the car-telephone industry] was a very brief blur [in the development of the mobile phone] because the guy who does any traveling—who goes to Philadelphia, Chicago, Boston and wants to get dial tone—can now take his phone with him in a briefcase."

You may have a sophisticated mobile telephone—but it's probably installed in your car. And unless your car is an extremely compact model, your car won't fit into an airplane or the trunk of a cab. That's why the briefcase phone was developed—because you *can* take it with you. The briefcase phone may well replace the car phone as the centerpiece of America's cellular fascination.

"The briefcase phones were introduced, not by major manufacturers, but by guys working out of garages. They've taken a $2 billion industry, put a gun to its head and pulled the trigger," Fisher said.

The briefcase phone can bridge the gap between the expensive cellular portable and the useful but mobile-only car unit. You can choose among several brands and types—and prices are considerably less expensive than the portable.

On a recent edition of the Diane Rehm talk show on WAMU-FM in Washington, D.C., a listener called in using her briefcase phone. "I want to add to some of the good news I've been hearing about cellular phones," she said. "The beauty of this briefcase unit is that it's upping my productivity with very little effort....I'm a sales representative for a paper production company, and I use it on the road and at the office. I've even taken it to the tennis court," she said.

Briefcase phone users are free from dependence on fixed phones or car phones. "If you had to draw a psychological profile of the typical user," Tony Amezquita of Liniair Mobile Communications in Denver said, "first, they would be very impatient—not the type that likes standing in lines. Second, they are very mobile between cities—they hop the plane and take their telephone with them." This ability to operate in both mobile and fixed settings, indoors and outdoors, is the principal benefit of briefcase communications.

The briefcase phone finds ready application in the construction industry. Installation of a conventional wired telephone on a construction site can cost thousands of dollars—as much as or more than the price of a briefcase unit. Unlike its wired counterpart, the briefcase phone can be moved rapidly from a construction trailer to an automobile and then to a new construction site.

Contractors who are not cellular subscribers but who instead use an SMR (specialized mobile radio) system can have SMR radios installed in briefcases to obtain dispatch and telephone communication on the job site or on the road. Specialized mobile radio offers you a combination of two-way dispatch communications between your office, vehicles and field locations, with the added advantage of built-in mobile telephone capability. The dominant companies in the SMR field are E.F. Johnson, General Electric, Motorola and Tactel.

In the event that your particular requirements are more specialized than either cellular or SMR, remember that almost any type of two-way communications equipment can be incorporated into a briefcase. One manufacturer, Liniair Mobile Communications, even constructs special UHF briefcase radio systems for ornithologists who track the flight patterns of birds.

The idea of a briefcase phone is not new. Units that operate on the older Improved Mobile Telephone Service (IMTS) mobile phone systems have been available for more than a decade, at prices under $4000. Although these units may be used while traveling through areas served by IMTS, they suffer from the intrinsic limitations of IMTS, such as long waiting periods to make a call due to frequency congestion.

It wasn't until the advent of cellular and SMR radio that the briefcase phone became popular—and practical—as a personal communications tool. Cellular and SMR systems offer higher quality service and more options than older types of mobile communications equipment.

That the briefcase phone has a ready market among those on the move, there is not doubt. As Walter Mitchell, account executive for Potamkin Communications in New York, explained, "Every one we've made, we've sold."

The briefcase phone typically contains a handset, transceiver, antenna, battery, charger and power cord. The unit can operate and be recharged from an auto's cigarette lighter. An AC power supply, sometimes called a "converter," can be used for operation near a wall socket.

Be aware of the difference between a "power supply" and a charger. The charger charges the unit's battery, but the power supply enables you to actually power the phone directly from a wall socket, without depleting battery power. Most units require you to open the case to use the phone. The OKI model has a flap that enables you to uncover just the handset portion when making a call.

A variety of antennas are available for briefcase phones. Some models have a variation of the "whip" or "pigtail" antenna used on vehicles. Some use internal antennas that are concealed inside the case lid.

An obvious advantage of the internal antenna is that your casephone appears in every respect like an ordinary briefcase. A case that must have its antenna extended in order to receive a call may not be as useful as one without this requirement. Ask before you buy and ensure that you will be able to receive calls with the briefcase closed if you desire that feature.

Metal, Leather or Plastic Case?

The trade-off between antenna type, case weight and case material may not be immediately obvious. The wood-and-leather or plastic cases are lightweight, and they have the added advantage that they can use an internal antenna. On the other hand, these cases do not offer as high a degree of protection for the phone as do the heavier metal cases.

"It's a challenge to build a case that will still work when you drop it and throw it into the trunk of a car," according to Andre Fortune, marketing director of Integrated Cellular Technologies (ICT) in Los Angeles. A metal briefcase is used in some models. The metal case is rugged, but it has a shielding effect and requires the use of the more obtrusive external antenna. A metal enclosure also adds weight.

Several manufacturers told us they are working on ways to reduce the weight of the cases, including redesign of the radio transceiver, which requires FCC approval. Current-generation briefcase phones weigh about 20 pounds. One working-prototype briefcase phone we saw weighed 10 pounds—and the production model will weigh 8 or 9 pounds, we're told.

Some briefcase phones feature an external antenna port. When using the unit in the car, you can connect this port to a whip antenna on the roof or trunk of the vehicle, providing better reception and transmission in many instances than if the briefcase phone were to use its own built-in antenna.

Most current-generation briefcase phones are essentially cellular car phones "repackaged" into briefcases. As a result, features that are standard on cellular phones such as memory dialing, electronic lock, scratch-pad memory and volume control are also available on briefcase phones.

You'll have to live without one standard feature of briefcases, however: room to keep things. Most of the room in the case is taken up by the radio transceiver, handset, battery, antenna and accessories. The inside lids of the cases do have some room for papers and slim files. As improvements in design reduce the size of the transceiver, more room will be freed up inside the case. Alternatively, smaller cases could be used. Perhaps we'll see a Dick Tracy lunchbox—with its own built-in cellular communicator.

The trend is to provide briefcase phones with special features not yet widely available on mobile phones. For example, the Telecase® from Integrated Cellular Technologies is available with either a 300-baud or 1200-baud built-in modem for data transmission.

"The modem-equipped model is going over extremely well in the medical field," according to ICT's Andre Fortune, "because it allows the physician to access the hospital's computer for scheduling information." Fortune emphasized that this type of case requires special attention to ensure that the modem works properly with the personal computer or data terminal that it is connected to.

An offshoot of the data transmission capability is facsimile transmission. Using a portable facsimile transceiver connected to the briefcase phone, you can send and receive diagrams, pictures, official forms and other types of visual or textual material. A portable cassette recorder will be available on future briefcase phones to enable you to tape conversations for future reference.

One manufacturer is even working on a hands-free option for the briefcase phone. This option will enable you to simply talk to the case—and have it activate and dial your calls, literally without lifting a finger. Such voice-recognition technology needs to be highly reliable. Otherwise, you might instruct your case to call Al Lee in New Jersey, only to find you've got Ali in Nairobi.

The most unusual feature we found in our survey of briefcase phones was "multiple roaming capability," available in the Western Union and Magnum-ROAMX™ products. Such a unit features multiple NAM (number assignment module) chips, making it possible for your dealer to "program" your phone to operate in several cellular systems without paying high roamer charges. A multiple-roam phone will appear as a "local" subscriber on any cellular system for which it is registered.

Of course, you would need to pay the regular monthly access charges in each system for cellular service as would any subscriber. If you regularly visit, say, New York, Chicago and Washington, this special roaming capability could keep you in touch while saving you money.

Because the central idea of the briefcase phone is communication away from wired phones or power mains, you will be dependent on the case's built-in battery. It is difficult to accurately determine how long the battery will last between charges because cellular is a full-duplex system—that is, you can talk and listen at the same time. To accomplish full-duplex communication, your cellular phone transmits and receives simultaneously, drawing more current than if it were to send and receive alternately.

How long the battery lasts depends on how much current the phone requires, how long the phone is actually in use making calls, how long it is listening for calls while idle and how long it spends charging. The cellular phones built into briefcases are selected for their low current requirements. Advances in design will probably lower the current requirements even further in future generations of the product.

Generally, you can expect to get eight hours of operation from your briefcase phone on one battery charge, including one or two hours of continuous talk time. Some models come equipped with a battery-status indicator to display the state of the battery's charge. To ensure that your phone has power available when you need it, plug it into your car's cigarette-lighter jack while in the car. The phone will charge itself from the vehicle's battery. While near an AC power socket, use the power supply or "converter" to power the unit directly from the mains.

Clearly, a phone that can travel with you—in and out of the vehicle, the office, the field—comes close to the ultimate personal communications device.

The go-anywhere phone—it's an open and shut case.

THE BRIEFCASE PHONE goes anywhere so that you never miss a call. Here Potamkin Communications account executive Walter Mitchell demonstrates a model assembled by his company, using an Oki phone.

THE CELLULAR CAR PHONE lets you turn driving time into office time. This phone is by NEC.

Chapter 14
How to Make the Most Efficient Use of Your Car Phone

The one thing to remember when purchasing your car phone is that you are not purchasing a piece of hardware, another gadget, a toy or a status symbol.

As we discussed in Chapter 4, what you are purchasing is *extra hours in your day.*

You will use your car phone to make the best use of the most precious resource in your life—*time.* The precise way you use the phone will, of course, depend on a variety of factors including the number of hours per day you spend in the car and the length of your commute to and from work.

One factor that you might overlook in calculating how valuable a car phone can be to you is the time zone you are in. The car phone owner in New York will use his car in a much different manner from the individual in Chicago or Phoenix or Los Angeles.

For example, the individual on the East Coast will use his morning drive time to talk with other people on the East Coast or possibly to talk to Midwesterners at home who are not yet on their way to the office. The East-Coast user may not be able to use his phone too much in the morning—unless he calls Europe frequently—simply because most people that he would talk to in his own time zone will also be commuting to work.

For the East Coaster, the most productive commute time for using his phone might be in the evening hours while driving home. From 5 to about 6:30 p.m. he can call the Midwest and the West Coast, where the people he would be calling would still be in the office.

The Midwest user would probably use his morning commute time to call the East Coast and his evening time to call the West Coast.

The West-Coast user would use his phone in the morning to call those people on the East Coast and in the Midwest who are already at the office. His homeward-bound hours would be less productive simply because few of the people he might call would still be at the office—unless he makes frequent calls to Japan.

Incoming and Outgoing Calls

You can, of course, use your phone in two ways: for outgoing calls and for incoming calls.

Surveys of the usage on the Chicago trial system indicated that the majority of mobile phone calls were placed by the mobile user from his car to a landline phone. A smaller but significant percentage of calls, however, did come to the mobile unit from landline phones.

For some occupations, it may even be more important for people to reach you than for you to reach them. For example, even though doctors may place few outgoing calls from their cars, they will need car phones so they can be reached in emergencies. The doctor may find it helpful both to carry a beeper and to have a car phone so that he or she can be reached both in and out of the car.

If you regularly communicate with people in other time zones, you could instruct them to call you during your morning or evening drive time at their convenience. Because you, rather than your secretary, will be answering your phone, they will be able to get through directly to you and thus minimize wasted time while they go through the usual ritual of waiting for your secretary to buzz you, confirm who's calling and so forth.

Also, they will pay the long-distance charges, which will keep your car-phone bill down. They may call from a time zone where the charges are lower, such as a person in Chicago calling you in Washington at 7:30 a.m. Central time while you are driving to work at 8:30 a.m. Eastern time. The Chicago caller would pay the before-8-a.m. charges, which are significantly lower than the after-8-a.m. charges for long distance service.

How to Make Your Car Phone an Extension on Your PBX

Picture the following: You are in your car, halfway between Washington and Baltimore. A call comes into your office in Annandale, Va., a suburb of Washington.

It's a call you've been waiting for for a week—from someone who you do not wish to give your car phone number to, yet you want to talk to the person.

Because of yet another marvel of modern technology, your secretary reaches out to you—in your car—and with the touch of a button "patches" the two of you together so that you can talk to the incoming caller. He need never know that you are in your car.

Most PBXs and advanced key systems include a function that allows instant conference calls.

Your secretary simply puts the caller on hold, dials you in your car and then patches the two of you together. She can even listen in and take notes as necessary during the conversation.

In essence, she will be treating your car phone as if it were an extension or another line on your PBX.

Call Diverting Increases Your Mobility

Furthermore, some of the more advanced PBXs allow what is called "direct inward dialing" (DID). By attaching your office line to a device called a "call diverter" (available in any phone store), a call coming in on your DID trunk line can be automatically diverted to your car phone. When you are in the office, the phone on your desk will ring. When you are in your car, your car phone will ring.

Most call diverters can be set remotely. You can call the diverter from wherever you happen to be and reset it so that the call is directly forwarded to that new number.

For example, if you are visiting a restaurant, you can request a table near a pay phone. You drop a quarter in the pay phone, call the diverter in your office and instruct it remotely to forward all calls that come into your office phone directly to the pay-phone number.

You could, of course, also divert calls—through its "call forwarding" function—directly from your *car* phone to the restaurant pay phone. If you use this latter method, however, you would have to pay the high air-time charges for cellular service even though the call itself ultimately terminates at a landline phone.

Before you leave the restaurant, be sure to put another quarter in the pay phone and once again forward your office calls to your car phone.

After Hours and Weekend Use

In most professions, little business is conducted after hours or on weekends. Even though you purchased your car phone for business use, you may find it of significant benefit in your private life as well.

A single man, for example, might use it to impress his dates.

You might be able to use it to help you discharge your obligations to your family. For example, you know you don't call your mother often enough. Call her from your car phone during those odd moments when you are stuck in traffic on the way to your Saturday date or on the way to the store.

Call when the rates are low. The fact that your consider her important enough to call from the car will impress her and be a big plus in your favor. She will realize—because you will remind her—that calling from a car phone is extremely expensive and, therefore, "I'm sorry, Mom, but this thing costs me 40 cents a minute, and I really can't talk too long."

You can get away with a 5-minute phone call and accomplish the work of a 20-minute call placed from home. Thus the car phone can help you manage your private time better.

Besides, any parent would prefer several 5-minute calls from his or her children at irregular intervals during the week rather than one lengthy "sense of duty" call placed grudingly once a month. The car phone may, thus, help to bring families closer together—a not insignificant effect in an era in which the nuclear bomb seems to be replacing the nuclear family as the major influence in our lives.

Chapter 15
Security: How Private is Cellular Telephone?

CELLULAR CALLS *offer a high degree of privacy. However, scanners such as the Regency Electronics MX7000 allow casual listeners to eavesdrop on conversations in the cellular band—although the likelihood that a listener will hear any one particular cellphone call is remote.*

Whenever I speak on the topic of cellular, I am frequently asked how private cellular telephone calls are.

Because radio waves can be monitored, no radio message is 100% secure.

However, compared to today's current car-telephone service, which is easily monitored by any Bearcat scanner, a cellular call is considerably more private—for several reasons.

First, cellular systems have not the 12 or 24 channels found in conventional IMTS systems, but several hundred—or even several thousand—different channels operating simultaneously in a city. Also, instead of a few hundred subscribers, a large city may have as many as 100,000 or more subscribers on the service.

The likelihood that anyone will be able to pick out your particular call from among the thousands on the air is highly remote. Furthermore, few scanners manufacturered today can pick up the 800 MHz band. These scanners are on the market and will come into more general use over the next several years. Also, the handoff process, in which your unit automatically jumps from one frequency to another as you drive from one cell to another, would make it difficult for the scanner to follow your signal. The listener may hear your conversation for only a few minutes before his monitoring is interrupted when a handoff occurs.

This frequency-hopping acts as an informal, low-level type of scrambling.

Second, to further protect your privacy, cellular companies may offer a more sophisticated level of scrambling as an option—probably at a slightly higher price per minute. This scrambling would be relatively unsophisticated, but it would be sufficient to foil the Bearcat-scanner crowd.

At some point in the future, the scanner manufacturerers will probably develop a sophisticated scanner that detects the handoff signal between the central MTSO and the cellular telephone on the road—which tells the mobile unit what channel to jump to—and will automatically make that jump at the same time the mobile unit makes the jump.

Third, your car's cellular telephone has a limited range. As you drive around the city, you will quickly drive out of the range of the average scanner listener.

In other words, even if the more-sophisticated type of scanner described above is developed, it will be able to follow your signal only so long as you remain in the range of that scanner. As explained in the first chapter of this book, because cellular uses low-powered transmitters, the signals are not broadcast too far beyond the edge of a cell. Therefore, the scanner will pick up signals only from those cells it is in close proximity to.

Fourth, the fact that there will be hundreds of thousands of mobile cellular phones in the city will also increase your sense of security. As the number of phones on the air increases, the likelihood of any one conversation being monitored drops dramatically.

The ability to monitor any one particular call can be extremely difficult. You may recall that it took scores of FBI agents and several million dollars of taxpayers' money to monitor all of the phone calls placed by Martin Luther King—and he was only one man.

Some ingenious detective agency will probably develop a program to monitor any individual's car telephone calls for a price. The number of man-hours involved in providing such a service may offer us a solution to our unemployment problems.

(Can you imagine the recruiting ad for an employee to work at this new car-telephone-conversation detection agency? "Be a James Bond! Ruin marriages! Learn the subtler forms of blackmail. Earn big dollars in your spare time. Send your alias and P.O. box number today for a free booklet.")

How Secure Are Your Telephone Calls Today?

How secure are your landline calls today? As Martin Luther King discovered, if a government agency wants to tap your phone, it can do so without too much difficulty, especially if large portions of its budget and its operations are secret. Even landline phone calls are not necessarily as private as they are commonly believed to be. *Caveat vocatus.* Caller beware.

Even if one of your conversations is occasionally monitored by a scanner owner who just enjoys listening to other people's phone conversations (where do people get the time to do things like this?), he will not know who he is listening to unless you idenify yourself over the air. Therefore, to assure maximum privacy, avoid giving specific identifying statements such as, "My home phone number is ————." or "My address is a ————." over your car phone if you are talking about any matters that you do not wish to be overheard.

The Digital Solution

All of the problems of lack of privacy associated with today's mobile telephones will be eliminated when "digital transmission" technologies come into common use over the next decade or so.

Digital transmission will present an even better way to scramble conversations. Cellular telephone—as currently formulated by the FCC rules—uses frequency modulation (FM), which is the technology that is also used on the standard FM broadcast band. FM provides a high quality signal that is not easily scrambled.

A signal transmitted digitally, on the other hand, would be ideal for scrambling because digital signals are easily coded and decoded by computer. The entire telecommunications network will probably be digital within a decade or two.

Chapter 16
The Alternatives to Cellular
You May Not Need a Car Phone

MILLICOM'S METAGRAM PAGER is a portable electronic mailbox. It allows you to receive complete alphanumeric messages anywhere within the receiving range of the system. The newer Metagram 900 (not shown) is even smaller and slips easily into the pocket or purse.

Cellular is the sexy newcomer to the mobile communications field. It is the personal communications technology that is getting all the press coverage and media attention because of the status appeal attached to having your own car telephone.

The cellular phone is the first mobile communications device that has been widely advertised and publicized since the CB (Citizens' Band) radio.

CB, of course, had its own problems, which we won't go into at this point. None of those problems affect cellular. Cellular technology does work and does perform its stated function quite well. As an attractive business tool, cellular will draw many people to the idea of mobile communications who never considered it before.

Despite the fact that cellular is the form of communications that is attracting most of the attention in the media, it is by no means the *only* personal communications technology.

Cellular, in fact, may not even be the answer to your specific telecommunications needs. A number of alternative mobile, portable and other services are available—often at a much lower price—that can do many jobs and may meet your needs quite adequately. These include paging, specialized mobile radio (SMR), two-way business radio and improved mobile telephone service (IMTS). You may also want to consider Citizens Band (CB) and amateur (ham) radio, which we won't discuss in this book. In addition, several other services are not yet available but should be coming on line within the next few years. These include digital transmission technologies and personal radio communication service (PRCS).

You should evaluate exactly what you need before you purchase your mobile-communications system.

In many cases, you may have to use an alternative to cellular simply because cellular is not yet available in your area or because you travel into areas where cellular is not available. Study the different services available and decide for yourself if they are adequate to meet your needs and do the job you need to have done.

Paging: Things that Go Beep in the Night

Paging has been around since the early 1950s. At one time only doctors and maintenance men carried pagers. That image persists to this day. Many executives still refuse to wear a pager simply because they believe the only people who wear pagers are the guys who fix the radiators and the plumbing in their buildings.

The pager can, in fact, be one of the most useful business tools available to the busy executive. Often all it takes to locate an executive when he is needed is a quick beep on his pager. He then finds the nearest phone and calls his secretary for whatever message awaits his action.

Because executives have refused to wear heavy beepers that clip onto the belt, however, the paging manufacturers have come up with a number of new pagers that they market as "executive pagers." These come in five types:

1. **The beeper** (sometimes called "tone only pagers"). Essentially, all a beeper does is emit some type of sound (usually a "beep") when its particular number is dialed on the telephone.

In other words, if you are wearing a pager with the number 555-1234 and your secretary wishes to reach you, she dials 555-1234 on the telephone. The paging system does the rest. It will automatically send your pager's identifying number out over the airwaves. Your pager is attuned to hear only its own signal. When that signal is broadcast, the pager will emit a beep to alert you that there is a message for you.

The early pagers were quite large and heavy and were worn on the hip. Some of the smaller pagers today are still worn on the hip. However, manufacturers have developed slim pagers that are about the size of a fountain pen and slip unobtrusively into the shirt pocket. Wristwatch pagers are expected to enter the market soon. Executives who resist carrying larger pagers that clip onto their belts may have less resistance to these new, tiny "executive pagers."

2. **Dual-address pagers.** More advanced pagers have "dual-address" capability. The dual-address pager has two telephone numbers and emits two different beeping sounds depending upon which of the two phone numbers is called. These are useful, for example, if one number is reserved to be called only when an emergency arises and the other is for more routine notification. Or, one beep could mean "call the office" and the other could mean "call home."

3. **Tone-and-voice pagers.** Tone-and-voice pagers offer a significant advantage over the standard beeper. The person placing the page picks up the phone and dials the pager number. At the sound of a tone, he has approximately six to eight seconds to deliver a quick voice message such as "John, please call the office" or "Steve, please call 555-7890 and talk to Mr. Jones." When the pager goes off, the recipient hears the actual voice message.

The disadvantage of the tone-and-voice pager is that sometimes it is difficult to hear what has been said because of background noise. You also may not have enough time between the initial beep—which warns you a message is about to come in—and the commencement of the message. As you scramble for a piece of paper to write the message down, you may miss the entire thing.

4. **The digital read-out pager.** Digital read-out pagers are quite useful devices if you are called frequently by a number of different people who are located in various places. The individual who wishes to reach you will pick up his phone and dial your pager number. At the sound of the tone, he uses the Touch-tone® pad on his telephone to dial his *own* telephone number. The paging transmitter will then transmit *that* phone number to your digital read-out pager. Your pager beeps and that phone number appears in a liquid crystal display (LCD) on your pager.

The digital-display pager can also be used to transmit messages in numeric form instead of phone numbers, according to prearranged codes. For example, the message on "4444" may mean "call your office," "6666" may mean "call your office immediately," "7777" may mean "you just won the state lottery and owe more money in taxes than you will collect in prizes" and so forth.

The digital display pager offers one major advantage over the alphanumeric pager, which we will discuss next: it can be addressed from any standard Touch-Tone® telephone. No special computer terminal is required.

5. **Alphanumeric (A-N) pager.** The most sophisticated type of pager available today is the alphanumeric pager, which probably should not even be called a "pager" at all because it does a lot more than merely page. Some manufacturers advertise it as a "portable electronic mailbox" or a "message center."

This portable electronic mailbox is capable of receiving complete messages in both letters and numbers. When your A-N pager goes off, it will deliver a complete message on its LCD display such as "George, Mr. Smith will be here at 2 p.m. instead of 1:30." You can even receive an A-N message anytime without disturbing anyone else simply by turning the beep off without turning the pager itself off—an option that most pagers offer.

Alphanumeric pagers are generally not inexpensive, but for the person who needs immediate access to information, they can be a godsend. They can be used with a number of timely information services such as stock market quotations. You can thus learn immediately that your favorite stock has dropped 5 points without waiting for the evening news or calling your broker.

The major disadvantage with the alphanumeric pager is that the message must be typed in on some type of transmitting computer before it can be sent to your portable mailbox. Some services provide toll-free numbers you can call to an operator, who will input the message for you. This input function may also be done in your office over your communicating word processor. Several manufacturers have also produced portable input devices that will work through any phone. With the numeric pager, on the other hand, you can be called by anyone who has access to a Touch-Tone® phone.

6. **The vibrating pager.** Pagers suffer from one problem that may get you thrown out of more nice places than you've ever been tossed out of before, namely the nasty habit the little things have of beeping in the middle of the quietest part of the symphony or during the silent prayer in church.

To get around this problem, some pagers have built-in vibrators. The vibrating pager allows you to receive a page without disturbing anyone.

Instead of beeping when it receives a message, the pager vibrates in such a way that you can feel it but those nearby are totally unaware that you have received a page. The vibration option is available with many different types of pagers, including tone-only, numeric and alphanumeric. With a vibrating alphanumeric pager, for example, you could receive messages in the middle of a meeting without anyone else knowing that you had been paged.

Pagers Are For Everyone

You should not think that the pager is for doctors and maintenance men only. Learn to think of it as another important tool that helps the executive better manage his time. Pagers are also becoming more accessible to everyone as their prices drop. Consumer pagers are available for less than $100 in many markets. That price should drop as the technology spreads.

You may, for example, give your child a pager that he can wear on the playground or to his friend's house. When it's time to come home for supper, beep him.

Expectant fathers have also made good use of pagers. I carried one during the last three months of my wife's pregnancy. The pager gave me peace of mind because I knew that if she went into labor, she could get in touch with me immediately—no matter where I was. I knew I would not miss the blessed event. (By the way, it was a boy. One of these days he'll probably be carrying his own pager around on the playground.)

Telephone Answering Machines

Telephone answering machines have been around for approximately 30 years. They have been called "the poor man's beeper" because they emit a beep that tells the caller when to begin leaving his message.

The answering machine is adequate to handle many—perhaps even most—business calls, particularly those that are not emergencies. Answering machines are fine for calls such as requests for information or orders from company salesmen. With a remote access device, you can retrieve your answering machine messages from any phone anywhere in the world.

Some answering machines will automatically dial your pager number after they take a message. In this way you will know that a message is waiting for you on your answering machine.

Voice Mail: A Computerized "Answering Machine"

(This section on voice mail is adapted from an article on the subject by Personal Communications Magazine assistant editor Elaine Lussier. The article appeared in the March 1984 issue of the magazine.)

> Voice mail is sometimes compared to a sophisticated-computerized form of telephone answering machine. Instead of messages being recorded on magnetic tape, voice-mail messages are recorded digitally and stored in a computer.
>
> Voice mail offers you a solution to the problem of the pink "while you were out" message slips that clutter your desk and your life. The pink slip problem can become almost unbearable for the executive who is out of the office for long periods of time.
>
> In effect, voice mail will turn your telephone into a combination of computer, answering machine and personal secretary. It will help you avoid the paper snowstorm and allow you to both receive and send accurate messages.
>
> The main advantage of voice mail as an executive time-saving tool is that it helps put an end to one of the biggest time wasters of all—the "dead end" telephone call, which is sometimes called "telephone tag." Dead end calls happen when you call someone and discover he or she is not available.
>
> How many times this week have you been on the "telephone merry-go-round"—the modern Olympic version of telephone tag?
>
> The merry-go-round begins when you call Bob Brown of XYZ Corp. and discover he is out of town. You ask his secretary to have him call you at his earliest convenience. When you return from lunch, your secretary hands you a pile of pink message slips. One of them bears the name and number of Bob Brown with the "returned your call" box checked.

Taking the matter (and the phone receiver) in hand, you call Bob's hotel in Kansas City—only to hear his phone ring and ring and ring—while you wait for the hotel operator to rescue you and take a message. Finally, late in the afternoon, your secretary informs you that, "Mr. Brown is on the line." You, of course, are on the other line, talking to an important client in Peoria.

Sooner or later you and Mr. Brown may meet by phone—if all goes well.

Observe how this scene would change if you and Bob Brown both had voice mail boxes:

You make the first call to Bob Brown at 10:30 in the morning. His secretary tells you he is out of town and asks if you would like to leave a message in his voice mail box. When you respond in the affirmative, she gives you his number, 5555.

(Note that the voice mailbox operates in a manner that is totally different from that of the telephone answering machine. Your call was answered by a live person—in this case, Bob's secretary—who switched you over to the mailbox. With the answering machine, you are answered by a recording. The request to leave a message in your voice mailbox may be presented either by a recording or by a live operator; the choice is up to you.)

Leave Your Message

After you have punched 5555 on your phone, you hear his message: "This is Bob Brown. I'll be in Kansas City until Thursday. If you would leave your name, number and a message, I'll get back to you as soon as I can...."

You greet Bob and announce that his order of 500 cellular phones is on its way. You tell him the address where they will be delivered and leave your phone number and voice mailbox number in case he wants to ask any questions. Confident he will receive the message the exact way you stated it, you hang up the phone.

Later in the day, when you return from lunch, the red "message waiting" light on your telephone is illuminated. You dial in your identification number and password, calling up the voice mail directory.

As you listen to the list of names of those who have called you, you notice that one of the messages is from Bob Brown. You dial up his message. He tells you of a change in the address for the shipment. You reroute Bob's shipment of phones to the address he has specified, and by 2 p.m. the shipment is out the door.

The result is that you have successfully defeated another round of telephone tag and perhaps saved an order that might have been lost or delayed.

Voice mail has proved to be the winner in every game of telephone tag and is quickly finding a home in the world of personal communications.

Studies have found that only one in four telephone calls reaches the intended party on the first try. If the person you call has voice mail, you wouldn't have to ask his secretary to have him return your call. Instead, you can leave a detailed message in his voice mail box and, in most cases, save a second phone call.

With voice mail added to your office telephone system, pager or cellular phone, you can say good-bye to missed or illegible messages. Say hello to accurate messages of any length, better communications with clients and associates, and improved productivity—not to mention the convenience of listening to your messages at any time and from any location.

The voice mail system digitizes the caller's voice and stores it in a computer. These messages can be retrieved from any Touch-Tone® telephone, thus making your phone messages only a push-button away.

Instead of hearing the traditional "I'm not here" message and the intimidating beep, the caller hears a personalized message that you have recorded in your own voice explaining that you are temporarily away from your phone and inviting him to leave a message. He is thus guided through the message-taking steps.

An operator's voice instructs him how to leave a message, review the message and, if necessary, make any changes in the final message. And rather than the customary 30 to 60 seconds or so allotted by answering machines, most voice mail systems provide as much as 3 to 6 minutes of message time.

An automated voice mail system also eliminates the misspelled names and other handwriting problems such as the "9" that looks like a "7." It also automatically inserts a date and time the message is received—a piece of information that may inadvertently be left off the pink slip.

Some callers might not want to tell a third party important corporate information and therefore won't leave a detailed message with a secretary. With voice mail, callers can supply one another with vital information—and do it in total secrecy.

Voice mail eliminates the middleman when busy executives attempt to contact one another.

Some people encounter a stage-fright-like condition when they have to talk to a machine. They forget what they wanted to say and stumble over words and ideas. With regular use, however, people can learn how to think and talk at the same time and leave comprehensible, complete messages.

Combined with Paging

Pagers and cellular phones are two ways to keep in contact while you're on the move. But what happens if your phone is in your car and you are in McDonald's when the most important call of the day comes in? If you subscribe to a voice mail service or if your cellular system provides voice mail, all you have to do is forward your calls to your voice mail box and allow it to take your messages.

A pager can provide instant notification of a waiting message. Subscribers to paging systems find that voice mail can answer their calls when they don't want to be disturbed.

THE DISTINGUISHED-LOOKING pen-sized Sensar pager by Motorola is an example of the new generation of "executive" pagers.

Electronic Mail

Electronic mail is a relatively new phenomenon. Electronic mail might be likened to the delivery of a telegram directly between two individuals. Telex, which has been around for several decades, is a type of electronic mail. Until the advent of the portable radio receiver, electronic mail has been limited to delivering messages between two or more individuals who were capable of retrieving these messages from a computer terminal tied into some type of wire such as a data circuit or a telephone link.

The alphanumeric pager, which we discussed earlier, is a type of portable electronic mailbox, which means that electronic mail messages can be delivered to you wherever you are—without wires.

In addition to its person-to-person capability, the alphanumeric pager also offers "point-to-multi-point" messaging service. This service is available through several new technologies including FM subcarriers and television sub-band carriers. These are too complex to explain in detail here. However, they do offer another type of alphanumeric paging.

For example, the Quotrek™ system provides stock-market reports on a real-time basis. Customers of the service carry portable receive-only data terminals that resemble calculators and print the latest stock quotes on a screen. The quotes are broadcast continuously over an FM subcarrier, a companion signal to the main signal of an FM radio station. Using a small keyboard on the unit, the user punches in the symbol for whatever stock he wishes a quote on. The unit captures only those quotes that the customer wishes to receive.

Conventional Two-Way Radio

Conventional two-way radio has been around for several decades and provides adequate—and in some cases superior—communications capability for fleet-type operations that are constantly under the control of a central dispatch location. Examples would be repair businesses, appliance-service outlets or plumbers.

Two-way radio has little or no status appeal, but it is useful in its place and provides an inexpensive type of two-way communication that can help a business carry out its functions. With repeaters, the service can be expanded to cover a wide area.

Specialized Mobile Radio

Specialized mobile radio (SMR) is a private radio service. (Cellular telephone is a common carrier service.) Changes in FCC regulations in 1982 allowed two-way radios operating on the SMR frequencies above 800 MHz to be interconnected with the telephone network and thus offer the option of both dispatch and telephone service from the same mobile radio.

SMRs are available in many communities. You will frequently see them advertised in the same pages of the local newspaper where you see cellular telephones advertised. The ads speak about car phone service at a low cost and avoid the word "cellular," but the casual reader may not realize the difference.

Before you sign up for SMR service, be sure to check the limitations that may be imposed on your use of your car radio as a mobile telephone. Your phone calls may be limited to one or two minutes. This limit may cause no problem if you are able to keep your calls from your car down to a minute or two, a trick that is easy to learn.

Definitely consider SMR if you have only occasional need for a car telephone but have a frequent need for dispatch service to a corporate fleet. The SMR radio combines both the dispatch and the mobile phone in one radio.

Wireless Data Communications

Communicating digitally by computer is considerably more efficient than voice communications. As much as 1,500 times as much information can be transmitted digitally as can be transmitted by voice, using the same amount of spectrum.

For example, speak the phrase, "Please return to the office immediately." That same phrase—typed onto a computer screen—in digital form could transmitted as many as 1,500 times using the same amount of radio spectrum that it would take to transmit the actual spoken words only one time.

Once the system is in place, two-way digital dispatch is considerably cheaper to use than voice dispatch. Federal Express could not perform its package pickup and delivery services without the computer-like two-way data terminals in every Federal Express truck. These terminals tell the driver where to pick up packages.

Some police departments also use mobile terminals in their police cars. If you remember *The Blues Brothers* movie, the policemen who stopped Dan Ackroyd ran his driver's license and car registration through a computer terminal located in the patrol car. The terminal was linked by radio to a mainframe computer back at police headquarters, which notified the policeman on the road that the car had an expired registration.

IBM has installed a system for its field representatives using wireless computer terminals made by Motorola. These terminals allow the service rep to check in with the host computer from anywhere in the local area. The system uses SMR frequencies.

The FBI also is reported to use some type of wireless mobile or portable communications terminals—reportedly made by Motorola—but the Bureau will, of course, deny everything if you ask, so don't bother.

Other organizations could make use of portable wireless two-way computers. Census takers and surveyors could use terminals to immediately put the material gathered into the host computer, avoiding the re-keying of survey forms. Portable terminals also would allow instant access to any of the 2,000-plus data bases that are available on line now and which are generally accessed through conventional wireline telephones coupled with computer terminals.

Currently most portable computers use 300 baud, which is a speed of 300 bits per second. That figure will be increased to 1,200 baud soon and as high as 4,800 baud in the not-to-distant future.

The portable terminal is also excellent for use in temporary locations such as construction sites, emergencies (such as during a hurricane), festivals that are in existence for only a short period of time, and election campaign headquarters.

Chapter 17
The Future of Mobile Communications

Where Do We Go From Here?

THE FUTURE is small and portable. You can get some idea of how small portable phones can be by examining the smallest commercially available today—a synthesized two-way radio made by Kenwood for the amateur radio market. The model displayed here by Tom Lott ("ham" call sign VE2AGF/G2CIN), a director of the Hallicrafters Co., lists for less than $250, has more than 800 selectable channels, weighs just over a half a pound and measures 2.24"×4.72"×1.1" in size—a truly pocket-size communicator.

No one knows precisely where this revolution in personal communications technology will take us. Could Alexander Graham Bell have envisioned the Mickey Mouse phone when he first said "Mr. Watson, come here I want you"?

When will we see Mickey Mouse cellular phones, with the antenna mounted in Mickey's ears, for the children? Will the executive who is afraid of being kidnapped wear a shoe with a secret transmitter built into the heel? Will fitness-minded salesmen close multimillion dollar deals over their wristwatch phones during their early morning jogs?

Most of us may not need such super-sophisticated gadgetry, but we would be delighted to have a telephone that we could clip onto our belt or slip into our shirt pocket or purse and use anywhere at anytime just as conveniently as the phone on the wall at home or on the desk in the office.

In large measure, the immediate future of this wireless personal-communications media will be determined by public policy regarding communications. If the cellular telephone industry is treated as a standard common-carrier type of service (such as the water company, electric company and the traditional telephone company) and is heavily regulated by the states, the rapid development of the industry may be stifled by nitpicking regulators.

On the other hand, if the regulatory agencies recognize that cellular is, in fact, a truly competitive service and that the pressures of the marketplace will protect the consumer, regulation will be unnecessary. The service will then be able to develop at its own pace as has the personal computer industry.

A variety of personal computers are available in all ranges of prices. No one suggests that we need to regulate Atari or Apple or IBM personal computers. The vagaries of the marketplace give the customer a wide range of choices and offer protection against "rip-offs" on the part of the manufacturers.

The FCC has made it clear that it has little intention of regulating cellular service and the other new personal communications technologies beyond the simple—almost routine—point of selecting the licensees and making sure that those licensees meet certain minimal requirements. Beyond that point the Commission is determined to let the marketplace govern. Such an attitude bodes well for the future of cellular and the other new mobile and portable communication technologies.

Many state public utility commissions also realize that cellular is competitive. A number of states have already abolished state regulation over cellular and other competitive radio common carrier services, and others are considering abolishing such regulation. Less regulation will mean that the power of the marketplace will help the service develop at its own pace and direction rather than in a direction artificially imposed upon it by non-market political forces.

A Bright Future

Technologically speaking, the future of this industry is bright. We can all look forward to the day of the Dick Tracy-style two-way wristwatch radio telephone—perhaps before the turn of the century.

How soon the day of the truly practical two-way wrist radiotelephone arrives will depend on two factors:

(1) The existence of a network of small cells. Low-powered wristwatch radios will work well only for a short distance; and

(2) The development of a much more sophisticated and advanced type of battery that will deliver higher power and/or longer life.

Small two-way radios the size of a deck of playing cards are already available for use in the amateur radio bands. They clip neatly onto the belt and will fit comfortably into a shirt pocket. Innovations that are first available for "hams" eventually find their way into the commercial market.

As of this writing several portable cellular phones are on the market and more are expected soon. These will also fit inside a man's breast coat pocket with some bulge.

Despite its great promise, today's 800 MHz cellular radio will probably never replace the conventional wireline telephone in most cities. With today's cellular technology, a cellular system can handle only about 100,000 to 200,000 customers on a single system—not nearly enough to serve the complete telephone communications needs of the larger cities.

Newer and better services, however, are now in the developmental stage that will bring efficient, inexpensive *wireless* communications capability to everyone. Two of the most promising new technologies now on the horizon appear to be amplitude compandored sideband (ACSB) and digital transmission. These will provide a more efficient use of the spectrum and allow more phone conversations to be made using the same amount of spectrum.

The personal portable communications device that you can carry everywhere and use all the time as a replacement for the wireline phone will eventually be the size of a wristwatch. I expect that it will use digital transmission and operate through a cellular-like land-based network and/or through direct communication to satellites.

As I see it, the wristwatch telephone of the future will contain a full miniature keyboard and combine two-way telephone with data capability, alphanumeric paging, standard radio and t.v., a voice-stress analyzer lie detector, a calculator, dictating machine and a word processor. It might even tell time.

No matter how this technology develops, the future will be exciting. John Naisbitt said in his concluding line to his book, *Megatrends,* "What a fantastic time to be alive!"

Equally exciting is the perception of the future expressed by the late Chester Gould, the creator of the comic strip "Dick Tracy" and the man whose vision helped set the personal communications revolution into action.

During an interview with this writer at Gould's farm in Woodstock, Illinois, in June 1984, the creator of the idea of the two-way wrist radio said:

"If you can think it up, it will be done."

Truer words were never spoken.

THE TELEPHONE OF THE FUTURE may eventually be wristwatch size. The Dynam Enterprises Wristalkie two-way communicator, modeled here by company engineer and designer Masashi Kuroyanagi, is primarily a toy with a limited range, but it does give a hint of what a future wristwatch phone may look like.

AFTERWORD

Cellular: The Beginning of a Revolution

The Social and Political Implications of Person-to-Person Global Telecommunications Will Play a Major Role in Your Future

by John Naisbitt

JOHN NAISBITT, the author of the best-seller Megatrends, *explains that cellular marks the beginning of the age of truly portable global communications.*

Editor's note: On April 2, 1984, Washington, D.C., became the first city in the United States—indeed, in the world—to have competing cellular systems.

The city's first cellular system, the non-wireline partnership Cellular One, had gone into service quietly and without ceremony on December 16, 1983, serving both Baltimore and Washington.

Meanwhile, Bell Atlantic Mobile Systems (BAMS) worked overtime to bring its competing wireline service on line. BAMS's system, "Alex" (as in "Alexander Graham Bell"), went fully commercial with four cells, serving part of the Washington area. The system's full 21 cells went into operation before the summer of 1984 serving both the Nation's Capital and Baltimore—thus giving Washingtonians and Baltimoreans full-fledged competition for their cellular business.

BAMS held a special press conference and ceremonies at the Vista International Hotel in Washington on April 10 to inaugurate the start of Alex service. A highlight of the ceremonies was a call over a cellular car phone from BAMS president Joseph Ambrozy to comedian Bob Hope. Hope was selected, Ambrozy explained, because he has shown in his decades of unselfish public service that he "personifies mobility."

The other highlight of the day was a keynote speech by John Naisbitt, the author of Megatrends, *which has sold more than 4 million copies. "[Naisbitt] properly sees what is happening to society and what is happening to this industry and is trying to awaken people to those trends," Amborzy said as he introduced Naisbitt.*

This afterward, which is based on Naisbitt's keynote speech, offers a fitting overview of the revolutionary impact that cellular telephone will have on our society as seen by one of the nation's foremost futurists.

*   *   *

Turning on a new cellular system here in Washington—the world's largest theme park—is part of the beginning of a revolution.

The introduction of cellular technology marks the beginning of the age of truly portable communications. Instead of connecting people via stationary telephones, cellular communications will connect the world on an *individual* basis. That is extraordinary.

Using the cellular phone in my car, I can call my colleague in Goteborg, Sweden, while he is driving around in his car. (At least I *tried* to do that, but I think he had his phone turned off when I tried it a couple of days ago!)

Just as the launching of the first satellites, the launch of cellular technology heralds a new stage in the development of the Information Age.

Within a few years, any person will have instant access to any information in the world no matter where he is on earth.

When the first Sputnik went up in 1957, people fancied that that was the beginning of space exploration. It was really the beginning of the *global information network*.

The U.S. space shuttles have done more to promote the global information economy in our lifetimes than they will ever do to promote space exploration.

Smaller and Smaller

Remember the first earth stations we had—the first satellite "dishes"? They had to be huge because the satellites contained low-powered transmitters that put out weak signals. The large dishes were needed to pick up that signal.

Now that whole process is being reversed. We are putting more powerful and more sophisticated satellites in the sky and will soon be launching satellites that are larger than a hotel ballroom.

The more powerful the satellite, the smaller the dish we need on the ground to receive the signal. We now have dishes that are only a foot or so across. Soon we will be able to work directly through the satellite with an antenna that is attached to a fully portable, handheld receiver.

I think that ultimately we will see satellites used to link cellular systems and bypass the landline networks completely.

Just as the launching of the early satellites in the late 1950s introduced the era of global satellite communications, the introduction of cellular communications will lead to another stage in the development of that global communications network.

Pocket-size Phones

The time is quickly approaching when we will be able to walk around carrying our telephones in our pocketbooks or even on our wrists. This is, after all, the century of miniaturization. There's already a cellular portable on the market that slips easily into the breast pocket of a man's suit coat.

In the next few years I expect we will progress from our cellular car telephones to a wristwatch-sized communicator that will give us access at any time to any information in databanks located in any part of the world.

The wrist telephone is something of a symbol. With the miniaturization and batteries available today, we already have highly portable telephones that we can carry around with us as easily as we carry a book.

It is also symbolic, I think, that cellular is starting out as a phone service in automobiles. For Americans, the automobile has always been a symbol of the freedom to pick up and go. Cellular adds another freedom on top of the freedom of movement the car gives us—the freedom to call and to receive calls while we are mobile.

An important consequence of the global communications network is that we are moving into a period in which we will not be dependent on any *national* communications systems.

We will be able to call anyone else who has a phone anywhere in the world—directly.

Breaking National Barriers

Think about that for a moment. In effect, this new multi-national satellite-based individual-to-individual communications system will break down the last of the national barriers.

That is extraordinarily important. People will be connected as individuals regardless of where they happen to be physically located at any one time.

In effect, cellular might help fashion us into some sort of global family. It will have social and political implications that are as profound as those which came with the introduction of the telephone itself to our society.

This entire development of a global communications network will be pushed by the extraordinary world economy that has been gathering momentum and is unfolding right now. When I speak of an international economy, I do not mean merely the increase of trade among 150 or so countries. The effect is far greater than that. What is happening is a shift from *trade among countries* to a *single unitary global economy*. We are moving toward that single global economy at a great rate.

Part of the development of that global economy is the interconnection of the more-than-one-billion landline telephones that are in place in the world today and the development of the direct-dial international telephone network.

Bypassing Governments

For example, I have friends in the Soviet Union. If I knew their telephone numbers, I could just dial them directly. Not only would the call not have to go through the rest of the system, but also the call would not be monitored because it would leapfrog the national system. The implications for totalitarian countries are just staggering.

Cellular may also bring telephones to people in Third World countries who have never had phones before. Cellular doesn't have to be mobile, of course. It can be stationary, too. In some more remote places people will have cellular telephones out on their back porches, powered by solar batteries.

Another factor driving this development is what I call "high tech/high-touch"—balancing what might otherwise be cold, impersonal technological wonders with the spiritual demands of human nature. This process is sometimes referred to as the development of "user friendly" machines. As they say in the phone company ads, "Reach out and touch someone." We'll be able to do it anywhere, anytime, with our portable telephones.

What will the effect be on us as individuals? It's still too early to tell, but I'm sure I'll want to get an unlisted number! I've already had a taste of those effects.

Love at First Dial

I was in London last week, and my car phone was installed in my car while I was away. When I got back on Saturday, I went for a ride and called everyone I could think of. I loved it!

I can't tell you how alarmed my office was when they discovered that I have a mobile phone.

I can bear testimony to how high the quality of these cellular phones is. Over the years I have used earlier models of radio telephones, and their quality hasn't been too good. But when I used the cellular phone, I was really struck by the quality—not only on calls to my office but even on calls to London.

I walk to work, but if I had a long commute, I'd certainly use it that way because commuting is such a waste of time. It would be less a waste of time if we could use that time not merely for getting from point A to point B but also to keep in touch with people.

Using a portable telephone to place calls from anywhere is a form of "collapsing the information float"—the amount of time that elapses between communications.

For example, if I send a letter to you, it usually takes three or four days to get to you. You may take another week to send me a reply. It will have take us perhaps 10 days to two weeks to negotiate something between us.

Now, if we communicated electronically, I could send my letter to you almost instantaneously. You may entertain it for an hour or two and then get back to me. We would have negotiated the entire transaction in not 10 days or 2 weeks but in an hour or two. We would have foreshortened the information float.

With the development of a global economy and a foreshortening—or collapsing—of the information float, we will have almost-instant access to each other. Whatever we negotiate, we will negotiate it much faster. The result is an acceleration of change.

Cellular phones will further contribute to the collapse of the information float. With a cellular phone, we'll be able to get something done instantly—wherever we are—and not have to wait until it's convenient for us to use a stationary phone.

What we are moving toward is a day in which we will be able to be in direct touch with anyone, anywhere in the world. We will dial him up on our wristwatch telephone. The signal will transmit directly to the satellite, and the satellite will be able to locate him—anywhere on earth.

It's all a product of the development of more powerful satellites and smaller and smaller dishes coupled with the birth of a global economy.

I think that all of these developments add up to nothing short of the beginnings of an important revolution that will change the world—and cellular radio will play an important role in that revolution.

Appendix

How To Make A Safe Call From Your Car

Make "Safety First" Your Motto

"**Safety first.**" That should be your motto—especially when you use your new car telephone.

If you are careful when you use it, your cellular telephone should present no problem to you, your passengers or other drivers on the road. Recognize that paying attention to your driving is your first and foremost responsibility while you are on the road. Never let your car phone interfere with your driving, and it will give you many years of profitable, *safe* use.

The following tips will help you make every call a safe call.

1. **Let another passenger in the car place the call** for you, if possible.

2. **Cut your driving speed while phoning.** Stay in the slow lane, and pay particular attention to traffic conditions. Before placing or answering a call, check the traffic situation to make sure you can use the phone safely. Keep your eyes on the road at all times.

3. **Dial while your car is not in motion,** such as while stopped at a traffic light or a stop sign, or pull over to the side of the road.

4. **Dial, pause, dial.** When dialing a seven- for 10-digit number (or longer), there's no need to dial the entire sequence of numbers at once. Dial only a few numbers, reassess the traffic situation, then return to dialing—and repeat this process until the complete number has been dialed.

5. Use the **"memory dialing" function** in your phone. This function allows you to record from nine to 99 phone numbers that you can recall instantly by touching only two or three buttons. For example, if you have accumulated a pile of "while you were out" pink slips from people you plan to call back while in the car, enter each individual's phone number in a separate memory-dialing location before you drive off. You can then quickly recall their numbers from the memory with a touch of only one or two buttons.

6. **Make sure that your phone is mounted where you can reach it easily** without having to move your body out of the position you normally assume while driving.

7. **Learn to operate the phone without looking at it.** Become familiar with the location of all the phone's controls so that you can press the various buttons without taking your eyes off the road. With a little practice, this process is easy.

8. **Tell the other party that you are calling from your car.** Explain that if a traffic problem occurs, you may have to drop your handset. The party on the other end of the line will then have some advance warning that you may have had to deal with an emergency.

9. **Use a hands-free phone, or speakerphone.** This will let you keep both hands on the wheel while you talk on the phone. Install the microphone on the visor directly above your line of vision. You can drive and speak without having to turn your head or take your eyes off the road.

10. **Use the phone's "electronic scratch pad."** If the person you are talking to during one call gives you a phone number to call later, the electronic scratch pad allows you to immediately record that number in your phone's memory so that you can dial it by punching only one one or two buttons. This feature is easier—and safer—than trying to write the number down on a piece of paper.

11. **If your call requires notetaking,** stop your car in a safe location or offer to return the call as soon as you can stop. If this is impractical, dictate your notes into a small tape recorder. Do not attempt to write while driving, which would require you to take your eyes off the road. Some manufacturers have promised to market car phones with a built-in dictation machine that can be used for this purpose. When this option becomes available, ask you cellular company to install it in your phone for you.

12. **Use automatic transmission.** Otherwise, you might find it impossible to hold the wheel with one hand, hold the telephone with a second hand and shift gears all at the same time. If your car has manual transmission, be sure that your phone offers hands-free service.

13. **Keep your calls brief.** This tip will not only help you drive safely but will also improve your ability to manage your time and keep your phone bill down.

14. **Practice using the phone while the car is stationary.** Become thoroughly familiar with all of its operations. Ask the dealer who installs the phone for you to give you a thorough demonstration of how it works. Read the user's manual.

15. **Make sure that the phone's handset is fastened securely** in its cradle when the phone is not being used. Otherwise it may fly around the car—and possibly injure you or a passenger—if you have to stop suddenly.

16. **Do not use your cellular phone near blasting caps or in an explosive atmosphere.** If you see a sign warning you to "Turn off two-way radios: blasting area," do so at once. Push the "off" button to deactivate the entire radio unit. Otherwise, the radio may transmit automatically if you receive an incoming phone call.

17. **Buckle up.** Use your seatbelt at all times while driving.

With a little practice and an awareness of the possible dangers of the mobile phone, you should be able to use your car phone safely. Carrying on a conversation over the phone should be no more difficult or unsafe than carrying on a conversation with someone seated inside the car—*if* you use caution.

Have a safe call—not a close one.

This appendix is based on an article that originally appeared in Personal Communications Report *(a newsletter for telecommuncations users), 4005 Williamsburg Ct., Fairfax, VA 22032. To subscribe ($24/year), call 1-800-227-2255 ext. 89. (In Virginia and the Washington, D.C., metropolitan area, call 703-352-1200)*

The Authors

Stuart Crump Jr. has been a professional journalist since 1972. He is the editor of *Personal Communications Magazine* and founding editor of *Cellular Radio News.*

He first became interested in mobile communications when reading the comic strip "Dick Tracy" while in elementary school. He subsequently became a radio amateur and currently holds a general class ham license with the call sign N4EGX. One of the highlights of his career was meeting and interviewing Chester Gould, the creator of "Dick Tracy." This was Gould's final interview; he passed on in May 1985. That interview is the cover story in the January 1985 edition of *Personal Communications Magazine.*

He has spoken on cellular radio at many national and international trade shows, including shows sponsored by the Cellular Telecommunications Industry Association, the Canadian Radio Common Carriers Association, the Eastern Management Group, Phillips Publishing, Telestrategies, the Tele-Communications Association, and the Organization for the Protection and Advancement of Small Telephone Companies.

He has also spoken at the Canadian Mobile Radio Communications User Conference, Mobile Communications Expo, Communications Network, Intelexpo, the International Conference on Consumer Electronics, the Federal Computer Conference, the Electronic Representatives Association's Repcom'84 conference, the Consumer Electronics Show, Speech Tech '85 and the Voice Input/Output Systems Application Conference.

He also chaired the American Management Associations seminar for communications executives, entitled "Mobile Communications Technologies: Facts and Formulas for Dollar-Wise Decision Making." That seminar provided the basis for this book.

Each December since 1982, FutureComm Publications has sponsored an annual conference on cellular and personal communications in Washington, D.C.

In addition to his regular features in *Personal Communications Magazine* and *Cellular Radio News,* his articles have appeared in most major communications-related trade publications including *Communications News, Telephone Engineer and Management, Telocator Magazine, Business Radio Action, Telephone Angles, Computer De-*

cisions, *Telecourier* and *Answer: The Professional Journal of the Telephone Answering Service Industry.*

He is a regular columnist for *Business Communication Review,* the magazine for telecommunication professionals, and a frequent contributor to *Mobile Phone News.*

He has also been quoted in major consumer and trade publications, including *The New York Times, The Washington Post, The Christian Science Monitor, USA Today, Esquire, The Washington Business Review, Science Digest, Management Technology, Financial World, U.S. News and World Report, Advertising Age* and *The Journal of Commerce.*

His previous experience includes 7 years (1972-79) with the *Princeton Packet* chain of weekly newspapers in central New Jersey, where he served as business editor. His weekly column won the National Newspaper Association's "best in nation" award. While serving as managing editor of one of the chain's newspapers, it won the New Jersey Newspaper Association's "best small weekly" award. Subsequently he became founding editor of *Telephone News* for Phillips Publishing Inc. (1980-81), which he left to found FutureComm Publications with his brother John in 1981. His address is: FutureComm Publications Inc., 4005 Williamsburg Ct., Fairfax, VA 22032, 703/352-1200.

Dr. Larry Baker, Ph.D., who wrote most of Chapter 4, has been president of Time Management Center, 3855 Lucas and Hunt, Suite 223, St. Louis, MO 63121, 314/385-1230, since 1978. The firm conducts seminars to show organizations and individuals how they can increase productivity by applying time-management techniques to daily problems.

Time Management's clients include Bell Laboratories, Monsanto Corp., Ernst and Whinney and General Motors Corp.

Dr. Baker's articles on organizational efficiency have appeared in a variety of publications.

He received his doctorate in business administration from Indiana University. Before joining Time Management Center, he was a professor of management at the University of Missouri at St. Louis. He is a member of the American Management Association, the Academy of Management, the American Psychological Association and the American Institute for Training and Development.

He is the author of three articles on using the car telephone as a time management tool. Part I, "The Cellular Mobile Telephone: A Luxury or a Necessity?" appeared in the May 1984 issue of *Personal*

Communications Magazine. Part II, entitled "How Much is Your Time Worth?" originally appeared in the July-August 1984 issue of the magazine. Chapter 4 in this book is based on those articles. Part III, entitled "Cellular Telephone: A Cost-Effective Time-Management Tool," appeared in September 1984.

Dr. Baker became acquainted with cellular telephone capabilities several years ago. As president of Time Management Center, he was asked by William Newport, then head of Bell's Advanced Mobile Phone Service (the AT&T cellular subsidiary at the time), to present several two-day time-management programs for the staff of AMPS.

Martin Cooper, who wrote the Introduction, is a consulting editor for *Personal Communications Magazine.* He was formerly the vice-president of Motorola Inc. and director of research and development. He is now chairman of his own cellular services software, information management and consulting firm, Cellular Business Systems Inc., 303 S. Northwest Hwy., Park Ridge, IL 60068, 312/698-5800.

He has been a pioneer, inventor, leader and authority in the mobile radio industry for more than 25 years. As general manager of the Communications Systems Division, he led Motorola's broad entry into the cellular radiotelephone business. Among his contributions are the development of the trunked system concept as used in SMRSs, the Pageboy II pagers and Metro terminals that heralded the beginning of high capacity common-carrier paging, and the IMTS mobile radio system.

He has been involved in the cellular business proceedings since their inception, and has spoken and written on this subject extensively. He introduced the first portable cellular telephone in 1973 and has strongly advocated portables since. He is a Fellow of the IEEE and of the Radio Club of America, is a recipient of the IEEE Centennial medal, has an M.S. from Illinois Institute of Technology and serves on various industry and technical committees.

Sue Easton, who wrote Chapter 12 on "How to Get the Maximum Status Value Out of Your New Car Phone," is the author of *Equal To The Task—How Working Women Are Managing in Corporate America* (Seaview, 1981). She is currently the editorial director of In Communications, 680 Beach St., Suite 349, San Francisco CA 94109,

415/441-1234, a print media public relations firm in San Francisco. Her free-lance magazine articles appear in a number of national periodicals.

Benn Kobb, who wrote the section in Chapter 13 on the briefcase phone, is the editor of *Personal Communications Report* and associate editor of *Personal Communications Magazine.* He is also an active radio amateur with the call sign KC5CW.

Elaine J. Lussier wrote the section in Chapter 16 on voice mail while she was new products editor for *Personal Communications Magazine* and *Cellular Radio News.* Previously, she was editor-in-chief (1982-83) of *The Southern,* a weekly newspaper at her college in Lakeland, Fla.

John Naisbitt, who wrote the Afterword, is the author of the bestseller *Megatrends,* (New York: Warner Books Inc., 1982). He is chairman of the Naisbitt Group, 1103 30th St. N.W., Suite 301, Washington, DC 20007, 202/333-3228, which specializes in social forecasting. He served on the White House staff in several capacities, including the post of special assistant to President Lyndon Johnson. He was senior vice president of the research firm Yankelovich, Skelly and White, and chairman of the board of the Center for Policy Process in Washington. He was chairman and president of the Urban Research Corporation in Chicago for 7 years. His 30 years of business experience include positions as an executive with IBM and Eastman Kodak.

Personal Communications Magazine, which began publication in May 1983, is published monthly by FutureComm Publications Inc., 4005 Williamsburg Ct., Fairfax, VA 22032, 703/352-1200. Subscriptions are $25 per year. The company also publishes the newsletter for telecommunications users, *Personal Communications Report* ($24/year). FutureComm also began regular publication of the first publication devoted to the cellular industry, the newsletter *Cellular Radio News,* in November 1981. That newsletter was sold to Phillips Publications Inc. in May 1985. Phillips combined it with *Mobile Phone News* ($287/year). To subscribe, call 1-800-227-2255 ext. 89.

Photo and Illustrations Credits

NovAtel, p.1; Bell Atlantic Mobile Systems, 6, 49; Time Management Center, St. Louis, 25, 28, 33; E.F. Johnson, 44; *USA Today,* 51; Ericsson Communications, 52; Magnum RoamX, 54; Sample bill for Cellufone prepared by Cellular Business Systems Inc., 59; Audiovox, 61; Oki Advanced Communications, 69; Glenayre, 70; Walker, 70; General Electric, 71; Panasonic, 71; Harris, 71; Blaupunkt, 72; Webcor, 72; Ameritech Mobile Communications, 76; Budget Rent a Car, 80; Darlene Rubin, TransMedia Consultants Inc., 87; Motorola, 96, 121; NEC, 98; Southern New England Telephone, 104; Regency Electronics, 109; Michael Campbell, 129; Stuart Crump Jr., 15, 20, 73, 75, 92, 103, 113, 125, 128.

Index

A

Advanced Mobile Phone Service (AMPS), 17
Airplanes, telephones on, 83-85
Alternatives to cellular, 4, 113-124
  See also: data communications, electronic mail, Improved Mobile Telephone System (IMTS), paging, personal radio, specialized mobile radio (SMR), telephone answering machines, two-way radio, voice mail
Amateur radio, 114, 125
Ameritech Mobile Communications, 18
Amplitude compandored sideband (ACSB), 127
Antennas, 74-75, 101
Audiovox, 61

B

Bargain, cellular is a, 20-43
Batteries, 94, 100, 127,
Bell Atlantic Mobile Systems, 6, 18-19
Bell Labs, 17
Blackwell, Roger, 5
Blaupunkt, 72
Briefcase phones, 98-104
Budget Rent a Car, 80, 81

C

Call diverting, 107
Call management, 105-108
Cellmaps™, 83
Cell sites, 11-14
Cellufone, 59
Cellular Business Systems Inc., 59
Cellular One, 18
Citizens Band (CB) radio, 114
Communications log, how to use one, 24-26
Competition, 18-19
Cooper, Martin, xi-xii, 95-96, 141,

D

Data communication, 70, 102, 123-134
"Dick Tracy," 101, 127-128, 139
Digital transmission, 111-112, 123-124, 127

E

E.F. Johnson, 44,100
Electronic mail, 113, 121-122
  Portable electronic mail box, 113, 116-117
Ericsson Communications, 52

F

Features, 61-72 (see also specific features)
Federal Bureau of Investigation, 111, 123
Federal Communications Commission (FCC), 15-17
Federal Express, 123
Fleets, equipping your company's, 81
Free service, 85-86
Frequency reuse, 13-14
Future, xii, 50-52, 125-128, 129-134

G

Gavin, Bob, 95
General Electric, 71
Gleason, Jackie, 91
Glenayre, 70
Gould, Chester (the creator of "Dick Tracy"), 128, 139

H

Hands-free speakerphone, 65-66, 78-79
Harris, 71
Hertz Corp., 81
History of cellular, 16-19
How cellular works, 2-3, 6-14, 110

I

IBM, 123
Illinois Bell, 17
Improved Mobile Telephone System (IMTS), 9-11, 82-83, 100, 110, 114
Information float, 52-53, 133-134
Integrated Cellular Technologies, 101-102

L

Land Mobile Radio Show, 9
Leasing a cellphone, 81
Liniair Mobile Communications, 99-100
Long distance calls, 105-106

M

Magnum-RoamX, 102
Millicom, 113
Mitsubishi, 54, 92
Motorola, 18, 93, 95-97, 100, 121, 123
Mounting the phone, 73-75

N

Naisbitt, John, 52-53, 128, 129-134, 142
National Car Rental, 81
NEC, 98, phone on back cover is made by NEC
NovAtel, 1

O

Oki Advanced Communications, front cover, 69, 100

P

Paging, 94-95, 114-117, 120
    Alphanumeric, 116-117
    Numeric, 116
    Tone only, 114-115
    Tone and voice, 115-116
    Vibrating pagers, 117
Panasonic, 71
PBXs, 106-107
Personal radio, 114
Politics of cellular, 15-19, 132-134
Portables, 92-104, 131-132
Potamkin Communications, 100, 103
Predictions, see Future
Price of service, 4-5, 41-43, 55-60
    Air time, 19, 22, 41-43, 56-58, 94
    Mobile units, 22, 50-52, 55, 56, 97
    Monthly service, 19, 22, 56
Privacy, 109-112
Professional Communication Systems, 99

R

Regency Electronics, 109
Rehm, Diane, 99
Renting a cellphone, 81
Roaming, 82-83, 93, 102

S

Safety, 63, 66, 76-79
Scanners, 109-111
Security (see Privacy)
Sommer, Elke, 87
Southern New England Telephone, 104
Specialized Mobile Radio (SMR), 100, 114, 122-123
Status, 3, 87-91

T

Tactel, 100
Telecourier magazine, 8
Telephone answering machines, 118
Time management, xi, xiii, 3-5
    Calculating the value of time, 27-34, 41-42
    Car phones as a time-management tool, 105-108
    How much is time worth? 20-43
    Time is money, 47-48
Time Management Center, 21-41, 140
Todd, Mike, 91
Transportable cellphones, 98
Two-way radios, 122

U

USA Today, 50-51
Use of cellular phones,
    by a banker, 95
    by a building contractor, 46
    by consumers, 40, 41, 107-108
    by a district manager, 34-38
    by a doctor, 47
    by a fast-food executive, 47
    by a glass salesman, 47
    by a housewife, 40
    by independent business owners, 41, 43
    by a lawyer, 46
    by parents, 40
    by a real estate agent, 46
    by sales personnel, 23, 38-40, 47-48
    by security personnel, 40
    by a trucker, 47
    on airplanes, 83-85

V

Voice-activated dialing, 78-79
Voice mail, 118-120

W

Walker Telecommunications Corp., 70
WAMU-FM, 99
Washington-Baltimore Cellular Telephone Co., 18-19
Webcor, 72
Where to buy a car phone, 58

OTHER POPULAR TAB BOOKS OF INTEREST

Transducer Fundamentals, with Projects (No. 1693—$14.50 paper; $19.95 hard)
Second Book of Easy-to-Build Electronic Projects (No. 1679—$13.50 paper; $17.95 hard)
Practical Microwave Oven Repair (No. 1667—$13.50 paper; $19.95 hard)
CMOS/TTL—A User's Guide with Projects (No. 1650—$13.50 paper; $19.95 hard)
Satellite Communications (No. 1632—$11.50 paper; $16.95 hard)
Build Your Own Laser, Phaser, Ion Ray Gun and Other Working Space-Age Projects (No. 1604—$15.50 paper; $24.95 hard)
Principles and Practice of Digital ICs and LEDs (No. 1577—$13.50 paper; $19.95 hard)
Understanding Electronics—2nd Edition (No. 1553—$9.95 paper; $15.95 hard)
Electronic Databook—3rd Edition (No. 1538—$17.50 paper; $24.95 hard)
Beginner's Guide to Reading Schematics (No. 1536—$9.25 paper; $14.95 hard)
Concepts of Digital Electronics (No. 1531—$11.50 paper; $17.95 hard)
Beginner's Guide to Electricity and Electrical Phenomena (No. 1507—$10.25 paper; $15.95 hard)
750 Practical Electronic Circuits (No. 1499—$14.95 paper; $21.95 hard)
Exploring Electricity and Electronics with Projects (No. 1497—$9.95 paper; $15.95 hard)
Video Electronics Technology (No. 1474—$11.50 paper; $15.95 hard)
Towers' International Transistor Selector—3rd Edition (No. 1416—$19.95 vinyl)
The Illustrated Dictionary of Electronics—2nd Edition (No. 1366—$18.95 paper; $26.95 hard)
49 Easy-To-Build Electronic Projects (No. 1337—$6.25 paper; $10.95 hard)
The Master Handbook of Telephones (No. 1316—$12.50 paper; $16.95 hard)
Giant Handbook of 222 Weekend Electronics Projects (No. 1265—$14.95 paper)
103 Projects for Electronics Experimenters (No. 1249—$11.50 paper)
The Complete Handbook of Videocassette Recorders—2nd Edition (No. 1211—$9.95 paper; $16.95 hard)

Introducing Cellular Communications: The New Mobile Telephone System (No. 1682—9.95 paper; $14.95 hard)
The Fiberoptics and Laser Handbook (No. 1671—$15.50 paper; $21.95 hard)
Power Supplies, Switching Regulators, Inverters and Converters (No. 1665—$15.50 paper; $21.95 hard)
Using Integrated Circuit Logic Devices (No. 1645—$15.50 paper; $21.95 hard)
Basic Transistor Course—2nd Edition (No. 1605—$13.50 paper; $19.95 hard)
The GIANT Book of Easy-to-Build Electronic Projects (No. 1599—$13.50 paper; $21.95 hard)
Music Synthesizers: A Manual of Design and Construction (No. 1565—$12.50 paper; $16.95 hard)
How to Design Circuits Using Semiconductors (No. 1543—$11.50 paper; $17.95 hard)
All About Telephones—2nd Edition (No. 1537—$11.50 paper; $16.95 hard)
The Complete Book of Oscilloscopes (No. 1532—$11.50 paper; $17.95 hard)
All About Home Satellite Television (No. 1519—$13.50 paper; $19.95 hard)
Maintaining and Repairing Videocassette Recorders (No. 1503—$15.50 paper; $21.95 hard)
The Build-It Book of Electronic Projects (No. 1498—$10.25 paper; $18.95 hard)
Video Cassette Recorders: Buying, Using and Maintaining (No. 1490—$8.25 paper; $14.95 hard)
The Beginner's Book of Electronic Music (No. 1438—$12.95 paper; $18.95 hard)
Build a Personal Earth Station for Worldwide Satellite TV Reception (No. 1409—$10.25 paper; $15.95 hard)
Basic Electronics Theory—with projects and experiments (No. 1338—$15.50 paper; $19.95 hard)
Electric Motor Test & Repair—3rd Edition (No. 1321—$7.25 paper; $13.95 hard)
The GIANT Handbook of Electronic Circuits (No. 1300—$19.95 paper)
Digital Electronics Troubleshooting (No. 1250—$12.50 paper)

TAB TAB BOOKS Inc.

Blue Ridge Summit, Pa. 17214

Send for FREE TAB Catalog describing over 750 current titles in print.